Women Military Pilots
of World War II

Women Military Pilots of World War II

A History with Biographies of American, British, Russian and German Aviators

LOIS K. MERRY

McFarland & Company, Inc., Publishers
Jefferson, North Carolina, and London

LIBRARY OF CONGRESS CATALOGUING-IN-PUBLICATION DATA

Merry, Lois K., 1947–
    Women military pilots of World War II : a history with biographies of American, British, Russian and German aviators / Lois K. Merry.
       p.   cm.
    Includes bibliographical references and index.

    ISBN 978-0-7864-4441-0
    softcover : 50# alkaline paper ∞

    1. World War, 1939–1945 — Aerial operations. 2. World War, 1939–1945 — Participation, Female. 3. Women air pilots — History — 20th century. 4. Women air pilots — Biography. 5. Air pilots, Military — History — 20th century. 6. Air pilots, Military — Biography. I. Title.
    D785.M47   2011
    940.54'4082 — dc22   2010038427

British Library cataloguing data are available

©2011 Lois K. Merry. All rights reserved

No part of this book may be reproduced or transmitted in any form or by any means, electronic or mechanical, including photocopying or recording, or by any information storage and retrieval system, without permission in writing from the publisher.

Cover image: Helen Richey sits in a cockpit before an air race, 1932 (Library of Congress)

Manufactured in the United States of America

McFarland & Company, Inc., Publishers
  Box 611, Jefferson, North Carolina 28640
  www.mcfarlandpub.com

This book is dedicated to the memories of Rudolf Klare, my father, and Theodore Merry, my father-in-law, both veterans of World War II. My dad was an aircraft armorer stationed in England. I remember his commenting once that he'd wait for the return of "his" pilot in "his" airplane. My father-in-law was an aviation machinist's mate stationed on the USS *Saratoga*. Through no fault of his own he could not complete flight training in Texas, so he missed his chance to become a pilot. I finally understand why those models of forties-era fighters and bombers graced the cabinets in my in-laws' kitchen.

Table of Contents

Acknowledgments	viii
Preface	1
Introduction	7

PART I: THE STORY OF THE WOMEN'S AVIATION UNITS

1 — Transition from Civilian to Military Aviation	11
2 — How Women's Units Came About	21
3 — Women's Flight Units	33
4 — Becoming Military Pilots	53
5 — Daily Work in England and America	77
6 — Hazards and Sacrifices	93
7 — War's End	114
8 — Conclusion	126

PART II: BIOGRAPHIES

9 — The Leaders' Stories	145
10 — The Pilots' Stories	156
Chapter Notes	195
Bibliography	205
Index	209

Acknowledgments

I am grateful to my colleagues at Keene State College for granting me sabbatical leave and awarding me faculty development funds to support my research in Texas. My fellow librarians, Kathleen Halverson, Deng Pan, and Margaret Barrett, assumed my faculty duties and supervisory responsibilities during my six-month sabbatical. My dean, Irene Herold, informed me of resources and encouraged me to pursue my scholarship. Interlibrary Loan supervisor Linda Madden deserves special thanks for diligently obtaining obscure materials from libraries throughout the country. Many other people offered words of encouragement throughout my writing project. You know who you are. I hope my book lives up to your expectations.

I am grateful to Kazimiera Jean Cottam, editor and translator of first-person accounts by the Soviet women, for her prompt and helpful response to my email requesting information on Marina Raskova. Dawn Letson, coordinator of Special Collections, Blagg-Huey Library, Texas Woman's University, wrote a letter in support of my sabbatical application and provided valuable assistance during my visit in 2008, including a tour of the archive. Ann Barton and Kimberly Johnson retrieved and copied materials from the Woman's Collection at TWU, and gave me helpful tips about the library Website. I thank them all for making me feel so welcome. Marianne Wood, director of the National WASP World War II Museum in Sweetwater, Texas, showed me the museum's archival materials, gave me helpful hints for my research, and led me to the film *WASPs and Witches* and the book *The Gremlins*. Needless to say, I was thrilled to visit Avenger Field.

Last, but certainly not least, my loving thanks go to my husband, Brian, who not only serves as my very best sounding board but also patiently and proudly supports my writing efforts.

Preface

Writing about World War II

Before I began writing this book, I thought of my topic, "women who flew military aircraft in World War II," as a nice neat package, neither too obscure to research nor too wide-ranging to manage. After all, there were relatively few women throughout the world who could even *fly* airplanes in the 1940s, much less fly *military* airplanes. It also seemed that there wasn't much that had been written either by or about them, especially when compared to the truly staggering amount of material on almost every other aspect of the Second World War. However, I discovered that investigating even a minor aspect of World War II is challenging, because every thread I followed was connected in some way to the entire fabric of that world-changing event.

Telling the story of the women pilots thoroughly and accurately meant investigating the events of a war that still seems to stretch out infinitely. More specifically, it involved learning about the social, military, and historical context of the 1940s. I also had to consider the personalities and motivations of the four women who became the leaders of those pilots, because they also affected the ultimate success of the women's units. The Second World War was global in impact, but it changed each individual who lived through it in a personal way. And the wartime situation in each country was, of course, unique. When I considered all of these elements, I began to realize how daunting my task would be. It seemed less and less possible for my book on women pilots in World War II to be comprehensive; so I decided instead to present an overview of the subject, featuring both commonalities and differences in the women's experience while paying particular attention to the four leaders who were responsible for forming women's aviation units.

Organizing their stories presented another challenge. The women pilots' experiences were similar in many ways, but their units' situations were unique. I wanted to address common themes across a wide geographical span, but still adhere to the war's chronology. Therefore, I incorporated accounts of the British, Soviet, and American women in each topical section, in a consistent sequence according to the dates their units were formed. Interestingly, their disbandment followed an exactly reverse order, and that is the sequence I use in the chapter covering the war's end. There is one exception to this general structure, due to the obvious differences between the Soviet women's experience in combat and the American and British women's experience working outside the combat zones. Therefore, the section about the Soviets' daily life appears in the chapter on hazards and sacrifices. The importance of their four leaders is the unifying thread weaving throughout the narrative. Their biographies constitute chapter nine, and selected biographies of women from Britain, America, the Soviet Union, and Germany make up chapter ten.

My Purpose

As far as I can determine, there are no books with a worldwide scope about women who flew military aircraft during World War II. I hope that mine helps to fill a gap in scholarship on this subject and brings the women pilots' experience greater prominence. Women flew military aircraft in support of the Allies and also the Axis, but, to the best of my knowledge, it was only in Great Britain, the Soviet Union, and the United States that women served in aviation units created specifically for them. Even so, their numbers were relatively low. Within the vast body of scholarship on every aspect of World War II, accounts of a small number of unlikely participants can languish unnoticed. As the war years slide into the ever more distant past, their stories might be overlooked, causing women currently engaged in military aviation to presume that they themselves are groundbreakers when in fact they are not. My book is an attempt to remedy that situation.

The Topic

I first learned that women had served as pilots in World War II while I was pursuing a graduate degree in library and information science at the University of Rhode Island in the early 1990s. Assigned a bibliography project for one course, I chose "women in the military" for my topic. That was how I discovered books about the American WASP. Their stories piqued my interest, so I read more widely. I found sources about women who flew for other countries as well, and I was surprised to learn that some of them had even flown in combat. However, I found many resource lists on the topic to be either incomplete or inaccurate, and sometimes both. In response, as I read more, I began assembling an annotated bibliography of my own. To conduct more extensive research, after I decided to write a book on the subject, I travelled to Texas Woman's University in Denton to use the Woman's Collection at the Blagg-Huey Library, the definitive WASP archive. I also visited the WASP Museum in Sweetwater, Texas, the location of Avenger Field where the WASP trained.

The Sources

My quest for information has been constrained by my need to find pertinent materials written in the English language. Therefore, the majority of sources I consulted concern the Women's Auxiliary Ferrying Squadron (WAFS) and the Women Airforce Service Pilots (WASP) in the United States or the Women's Section of the Air Transport Auxiliary (ATA) in the United Kingdom. I discovered sufficient English language materials about the Soviet women to support my narrative, but I found relatively little information about women pilots in Germany. I found no indication that women actually flew airplanes in the Women's Auxiliary Australian Air Force (WAAAF), although the many thousands of women who served in that organization were known — somewhat inaccurately — as "airwomen." Two Australian women flew in the Women's Section of the ATA. Canadian women served in either the ATA or the WASP because, for them, no possibility existed for flying in the Royal Canadian Air Force (RCAF).

Materials about the women pilots vary widely in both character and quality. There are many excellent sources, some as riveting as any adventure novel. Others are of lesser quality. Some frustratingly omit important, even key, facts, names, locations, or dates. Depending on the women's narrative skill, their autobiographies are either a wonderful reading experience or nearly impossible to decipher, with texts that are disjointed and have no index. Nevertheless, the women's autobiographies serve as valuable first-person accounts.

I encountered a few earnest champions of feminism whose passion impedes their view of a broad and complex picture. Some of their books lapse into outright hero worship. One writer expressed disappointment that the WASP director, Jacqueline Cochran, never behaved as a feminist role model should. In my opinion, to require a woman to become a self-conscious role model only imposes other, more subtle, restrictions on her. All the evidence I've found shows that Cochran was truly her own woman, and, as such, she unconsciously modeled a different ideal; she was a person with no need to champion a cause because she remained at all times freely and unapologetically herself. Furthermore, current societal values should never be applied to those who lived in a different time. Doing so is an error of historiography, in my opinion—a pitfall I have tried to avoid by remembering that social context is a compelling influence on behavior.

Research Challenges

Determining the accuracy of some of the accounts is difficult. One book, Myles' *Night Witches: The Untold Story of Soviet Women in Combat*, about the Soviet night bombers, called "night witches" by the Germans, apparently stretches the truth a lot. My suspicions about the book's veracity came about when I read further and noticed that a few of the most vivid stories in *Night Witches* fail to appear in subsequent, more detailed versions of the same events in other sources.

Autobiographical sources are colored by the subject's personality, point of view, or political stance. *Flying Is My Life*, by the German test pilot Hanna Reitsch, is a gripping and illuminating life story, but the reader must allow for Reitsch's intense and unquestioning patriotism. Jacqueline Cochran, it turns out, crafted her own legend from fragments of her life story, after she left her birthplace behind, so anyone reading her autobiography, *The Stars at Noon*, must acknowledge that it is one more step in that effort, reflecting as it does Cochran's self-made persona, a blend of fiction and fact. In much the same mode, *The Heroic Flight of the "Rodina,"* although not an autobiography, is typical of official Soviet era narratives. It is the record of the long-distance flight made by a three-woman crew, including Marina Raskova, founder of the Soviet women's aviation regiments. Yet, despite the book's obvious hyperbole, the amazing story of the women's courage and endurance shines through.

If the woman died, either during the war or soon after, there is much less likelihood that a personal record of her wartime activities exists. Marina Raskova was killed flying an airplane in a snowstorm to join her regiment, already at the battlefront, in January 1943. Her memoir, *Zapiski Shturmana* (*Notes of a Navigator*), written after the *Rodina* flight, but before the war, has not yet been translated into English.[1] Pauline Gower, who founded the Women's Section of the ATA, died early in 1947, shortly after giving birth to twin boys. Gower wrote a very entertaining account of her "joyriding" years,

Women with Wings, so her postwar memoirs would have been both interesting and informative. The 1995 biography *A Harvest of Memories: The Life of Pauline Gower, M.B.E.*, compiled by her son, Michael Fahie, and based on the reminiscences of Gower's contemporaries and original documents from her estate, is the best source of information about her life.

The paucity of sources on German women pilots is understandable. Victors of war don't hesitate to record their histories, so American, British, and to a lesser extent, Soviet accounts of World War II abound; but, even without the language barrier, it is difficult to tease out information on the activities of German women pilots. One exception is the fairly extensive collection of materials available on Hanna Reitsch, a skilled pilot who was commissioned a *Flugkapitan* (flight captain) in the Luftwaffe, and who performed exemplary flight-testing work on behalf of the Nazis. Reitsch's rival in German military aviation, Melitta Schiller, Countess von Stauffenberg, was equally skilled, but has received relatively little attention, perhaps because she perpetually inhabits the shadow cast by her husband, Alexander, and his brothers, especially Claus, the moving force behind the planned coup d'etat and failed assassination attempt against Adolf Hitler in 1944.[2]

Although the Soviet Union was among the victors of The Great Patriotic War, the repressive nature of Joseph Stalin's government, which persisted into the postwar era, inhibited more than a few of the former women pilots, making them reluctant to discuss their wartime experiences, even if permission could be obtained to interview them. Because of the difficulties involved in gaining access to these women's stories before the collapse of the Soviet Union, most accounts of their activities have been available only since 1990.

An advantage to writing a book about the Second World War so many years after that war ended is that more materials are published on the subject all the time, and I've been fortunate to have had new information available. One recent biography, *Jackie Cochran: Pilot in the Fastest Lane*, upends earlier biographical accounts of the WASP leader because of its author's discoveries concerning Cochran's early life, the result of Rich's thorough search in public records. Another new biography, Rickman's *Nancy Love and the WASP Ferry Pilots of World War II*, fills in significant gaps in scholarship about the leader of the WAFS. Rickman interviewed Love's three daughters and gained access to her personal papers. A 2009 translation of Anna Timofeyeva Yegorova's *Red Sky, Black Death: A Soviet Woman Pilot's Memoir of the Eastern Front* gave me access to the life story of a Soviet woman pilot. Timofeyeva Yegorova served as the only woman in her regiment during the war. Translators Margarita Ponomaryova and Kim Green have produced an excellent and highly informative text.

There are numerous websites about women pilots; and although much of the information they present is both accurate and useful, it is equally likely that their content is erroneous and incomplete. It was the "hit or miss" nature of resource lists on those websites that initially spurred me to compile an annotated bibliography of my own. For scholarly purposes, websites are not especially reliable, but I have used those that I consider to be worthy resources.

Introduction

In the great cataclysm known as World War II, or for the Soviets, the Great Patriotic War, customary constraints on the assigned roles of women loosened and they were accorded opportunities for employment previously unknown. One of those opportunities was the chance to fly military aircraft. It is a little known fact that throughout the world at that time, thousands of women flew every type of military aircraft then in existence, including the most technologically advanced jets. As pilots, they performed mundane tasks such as ferrying aircraft from factories to air bases or transporting military personnel. But they also engaged in far more dangerous work, like testing new and experimental aircraft, delivering war-weary airplanes to their final destinations, towing targets behind airplanes for live ammunition practice, and other kinds of combat training or flying fighters and bombers in combat. They faced many hazards, poor weather conditions being the most persistent danger. A significant number of them died during training or while performing their routine work. The health of others suffered permanent damage. Paradoxically, for most of them, their service as pilots of military aircraft during the war years was a never-to-be-forgotten, and life changing, experience.

Integrating women pilots into military aviation was usually an incremental process in the countries where women's units materialized. In some places the very idea of female air force pilots was rejected outright. Individual women seemed unable to pursue military service as pilots on their own, although rare exceptions existed, such as Hanna Reitsch and Melitta Schiller in Germany and the Australian Nancy Lyle, who was very likely the only woman pilot serving Australia's armed forces, having volunteered to fly her Hornet Moth to train Australian Air Force ground forces in anti-aircraft maneuvers.[1]

When it happened that the participation of women was considered beneficial enough to allow for a temporary suspension of established roles, in favor of new untested ones, the time was ripe for the idea of women flying military airplanes. These women varied in background, temperament, and motivation for joining the units, except for an almost universal passion for flying. Likewise, the unique circumstances in their home countries determined whether or not they would be allowed to fly in combat. Cultural prohibitions against women pursuing previously all-male endeavors, especially flying military aircraft, were extremely powerful. Although some governments temporarily ignored those prohibitions, so as to increase the number of participants in the war effort, others, as in Hitler's Germany, resolutely held to cultural expectations and never officially allowed women such latitude.

No matter how they were employed during the war, one common theme throughout the world was that the presence of female pilots on military air bases was unprecedented and therefore no models existed, for either the women themselves or their leaders, to draw upon. Their arrival on air bases often caused consternation among their male colleagues, and they met resistance in their quest for full acceptance as military pilots. Whether serving their countries as civilians (as in Great Britain and the United States) or as full members of the army (as in the Soviet Union), women's flying units made the best of life in a context that previously had had no need whatsoever to accommodate them. For many, it turned out that the doors, which opened wide during the war years, closed almost as tightly as ever at the war's end.

The most compelling determinant of whether women could be military pilots during the Second World War was the level of peril perceived by their home countries, but another factor may be equally important — the existence of a well-known and charismatic champion who could promote the idea of employing women in uncharacteristic roles and who would doggedly persevere toward that goal in the face of strong opposition. Pauline Gower in Great Britain, Marina Raskova in the Soviet Union, and two pilots, Jacqueline Cochran and Nancy Love, in the United States, were such women. Without them, it is unlikely that there ever would have been officially sanctioned units of women who could fly military aircraft. Gower founded the Women's Section of the Air Transport Auxiliary (ATA) in Great Britain; Raskova formed Aviation Group #122, which was subsequently divided into three women's regiments in the Soviet Union; Love assembled the Women's Auxiliary Ferrying Squadron (WAFS), and Cochran initiated the Women's Flying Training Detachment (WFTD), which in time absorbed the WAFS to become the Women Airforce Service Pilots (WASP) in the United States.

Part I

*The Story of the
Women's Aviation Units*

1

Transition from Civilian to Military Aviation

Between the World Wars

It has never been common practice for women to be military pilots anywhere in the world. Shortly after the aviation era dawned, a woman from Texas, Katherine Stinson, the "Flying Schoolgirl," became the fourth woman in America to receive her pilot's license. Stinson racked up a number of "firsts" as a female pilot—first woman to perform a loop; first woman authorized to carry the U.S. mail; first person to fly an airplane at night.[1] When the United States entered the First World War, Stinson wanted to use her flying expertise to help her country's war effort. She volunteered to fly combat missions for the United States Army, but she was rejected—twice—because she was a woman. Another American pilot of that era, Ruth Law, also wanted to fly in combat, but was not allowed to do so.[2]

Airplanes were used for reconnaissance and for battle in the First World War. When the war ended, they became increasingly common and during the interwar years, they were put to civilian uses. In the Great Depression of the 1930s in the United States, airplanes were first used for mail delivery, initially by military pilots and later by private contractors. Most people got their first exposure to airplanes when they attended air circuses, shows featuring daredevil pilots, many of them veterans of World War I, who thrilled the crowds below by performing risky stunts in light airplanes. Joining these early joyriders in the skies over England was the daughter of a British member of Parliament, Pauline Gower.

Joyriding

Pauline Gower

Pauline Gower followed a path contrary to common expectations for women born into her social class. Her father, Robert, wanted his two daughters to be well educated, so Gower attended boarding schools in England and a finishing school on the Continent. However she grew bored at school and she was anxious to finish her formal education and pursue her interest in aviation. She joined the air club at London's Stag Lane, where she met the woman who would become her business partner, Dorothy Spicer. A photo from the flying club at Stag Lane depicts Gower and Spicer in coveralls hovering over a dismantled aircraft engine. The two friends formed a business, Air Trips, a joyriding and air-taxi service. Gower, the pilot, took onlookers for short rides and transported other passengers, while Spicer, the flight engineer, serviced their tiny biplane. The women operated the company from 1931 to 1936. By 1934, they had provided more than 10,000 passenger flights.[3]

Air circuses roamed the skies over England in summer's good flying weather, doing stunts and giving rides to anyone willing to pay a few shillings for some exciting minutes in the air. It wasn't a comfortable way to earn a living because it involved nearly constant travel and required the performers to camp at countless local airfields, wherever they thought new customers could be found. After Air Trips ceased operations, Gower wrote about her experience in *Women with Wings*. Describing a typical day, she writes:

> Early morning sees the engineers busy inspecting the aeroplanes, the pilots studying their maps, and the rest of the staff hurriedly striking camp. Then comes the flight to the next place on the itinerary. Weather may be bad, but still, if it is humanly possible, we get through it, exerting every effort to arrive at the new town in time to make a fresh camp and settle in before the afternoon show begins.[4]

In double-winged, open-cockpit airplanes constructed of wood, canvas and wire, air circus pilots performed tricks designed to satisfy the public's demand for novelty. The air circus began with a "prop," or propaganda, flight in which the airplanes flew in formation to advertise the show's arrival. Stunts included exhibition flying, air races, crazy flying, and tricks like "bottle shooting from the air." For that trick, the plane's engine was temporarily shut off so the crowd could hear shots fired from the pilot's blank pistol, while a circus worker, concealed behind a screen on the ground, waited for the right moment to shatter the targeted bottle with a hammer. Gower writes that she carried parachutists, including a "birdman" who wore silk bat-type wings with which he "flew" briefly before finishing the downward journey by parachute. Harkening back to the airplanes' function in wartime, one trick consisted of bombing a speedboat, or, farther from the coast, an automobile, with sacks of flour. One day a passenger (seated in front of her) hurled a bag of confetti onto a honeymooning couple in a speedboat on the water below, an action that also filled Gower's mouth with bits of paper.

1—Transition from Civilian to Military Aviation

Pauline Gower wears flight gear in a 1937 portrait. Her company, Air Trips, the only joyriding and air taxi service established and operated by women, served thousands of customers in the 1930s. Gower later founded the Women's Section of Britain's Air Transport Auxiliary (© National Portrait Gallery, London).

Their passengers' ignorance and excitability made them a potential danger to airplane and pilot alike. Gower explains:

> Small children are often a source of worry to a joyriding pilot. They seem determined to put their feet through the wings. I dislike intensely taking children up for more than a very few minutes unless I have had them under strict supervision for some time previously, for after having wrapped themselves round a dozen ices and lollipops prior to their flight, they usually feel ill after a very short period in the air, especially if it is a bumpy day.[5]

Engaged in a highly unusual occupation for females, Gower and Spicer were not immune to criticism, including negative comments from members of their own sex. She continues:

> The same week during a lull in the programme we went into a near-by café for some tea. Just as we entered the doorway two old ladies got up and, walking straight to us, took one look at our overalls and said,
> "Modern womanhood — disgusting!"
> This was too much for me.
> "Madam," I asked, "would you prefer us to fly in skirts, which would most assuredly end up blowing round our necks?"
> But not deigning to reply, they gave us another look and marched out.[6]

Early airports followed only rudimentary safety precautions. Gower tells the story of a local woman, leading a dog and carrying a baby, who insisted on claiming her legal right to get to her home, on the opposite side of a runway, by crossing while an air show was in progress. Despite the concerted efforts of air circus workers, which included grabbing the howling baby and seizing the dog, the woman could not be convinced to go around!

Setting Aviation Records

Marina Raskova

Besides their joyriding, pilots in the interwar period competed for records in distance or speed, and many aspired to fly farthest, highest, or longest, or to achieve other aviation "firsts." Soviet aviation technology took on the challenge of connecting far-flung settlements in a vast country. Setting long-distance records was part of that effort. The Soviet government officially encouraged women to learn to fly. In the 1930s, routine travel by air was mostly a dream of the future; but that dream came closer to reality thanks to the efforts of women like Marina Raskova, a citizen of Moscow and a Party member who had chosen to study science rather than pursue a career in music because she thought her chances of supporting herself as a scientist would be better. Raskova flew in air shows and races and she made several long distance flights, one of which involved flying through four different types of air masses — tropical, continental, polar, and arctic — each requiring special flying skills. Raskova was the first female navigator in the Soviet Union. She was navigator in a crew of three women who made aviation history in September 1938 when they flew a strikingly modern-looking airplane featuring the latest in Russian technology, an ANT-37 named *Rodina* (*Homeland*), from Moscow

Marina Raskova in uniform in 1936. She was one of three women who set a nonstop distance record by flying an ANT-37 named Rodina *across the Soviet Union in 1938. Raskova was as well known in the Soviet Union as Amelia Earhart in America (courtesy Library of Congress).*

across the country nearly to the Pacific, in the Soviet Far East, setting a new women's nonstop distance record of 6450 km.[7]

What made their flight so remarkable was the unanticipated event of Raskova's solo trek through the wilderness at the end of their journey. Anticipating an emergency landing because the fuel level was too low for the airplane to reach their planned landing site, the pilot ordered Raskova to bail out from her navigator's cockpit in its vulnerable location in the nose of the airplane. Landing in the rough terrain below could have upended the *Rodina*, so Raskova jumped, immediately finding herself alone in the uninhabited subarctic conifer forest, called *taiga*, with only the few provisions she had stowed in her pockets, along with a firearm and compass. Her comrades fired several shots to signal her, but their echoes were even louder than the shots themselves, leading Raskova away from their landing site. She realized her error only several days later and then changed her course.

There was plenty of water in the wet terrain; and after her supply of chocolate and mints ran out, she survived on mushrooms, cranberries and birch leaves. It took her ten days to reach the *Rodina*, which was intact enough to be flown out later, after the ground froze, having landed with its retractable wheels up. Soviet officials, who had been following the women's progress by radio, launched a massive rescue effort. Even after the three women reunited at the crash site, because of poor weather conditions it was several days before rescuers located their aircraft from above. Had there been snow cover, the light-colored *Rodina* might not have been spotted. The first leg of the women's return trip to Moscow was by canoe through the wilderness, and then back across the country by rail. They were welcomed as heroes all along their route. A rhapsodic account of the women's accomplishment, titled *The Heroic Flight of the "Rodina,"* was published shortly thereafter. Great significance was attached to the fact that the ANT-37 was of entirely Soviet design and manufacture, although the *Rodina* was in reality the only one of four prototypes that was operational. The women's flight pared travel time from border to border in the Soviet Union to 26 hours and 29 minutes, an amazing speed for the day.[8]

Jacqueline Cochran

In the United States, Jacqueline Cochran, who grew up in Florida logging camps, became a beautician who developed her own line of cosmetics. Cochran took her first flight lesson in response to a challenge posed by her future husband, wealthy businessman Floyd Odlum. Odlum told her that learning to fly would be the best way to keep track of her cosmetics business, but she was reluctant. He wagered that she could not learn to fly in three weeks, but Cochran soloed and passed her flight tests in only seventeen days, winning the bet, and simultaneously discovering her life's passion.[9] By that time, the new pilot could afford her own airplane and she soon purchased one. She entered air races and either won or placed second, but some of the air races were closed to female pilots. Cochran and her friend Amelia Earhart lobbied officials of America's prestigious Bendix Transcontinental Air Race to allow women to compete. After Earhart's attempt at an around-the-world flight ended with her disappearance somewhere over the Pacific Ocean, Cochran won the Bendix Trophy in 1938, flying an experimental long-range Seversky AP-7 pursuit plane. Jacqueline Cochran set a women's altitude record in 1939 and two absolute speed records in 1940.[10]

Air Marking, Test Piloting

Nancy Love

Nancy Harkness Love, a physician's daughter who grew up in Michigan, learned to fly in 1930; and, while still a student at Vassar College, she was occasionally disciplined for "buzzing" the campus. Love left college before graduating and eventually found work in Boston. She entered a few air races too, but never enjoyed that kind of flying. Her aviation pursuits were of a more pragmatic nature. In September 1935, Love began working with the United States Bureau of Air Commerce as a pilot in the National Air Marking Program, a scheme designed to mark roofs and other highly visible spots at the intersections of 15-mile-square grids in 16,000 American cities and towns, with the goal of orienting the pilots who were then flying throughout the country.[11]

Love also test piloted a new invention — the aircar. In 1937 she became a test pilot for the Gwinn Aircar Company of Buffalo, New York.[12] There she flew the prototype of an airplane/automobile which was designed to sustain a full perpendicular drop from 2000 feet, its three-wheeled undercarriage absorbing the entire stress of impact. Love's job was to generate data for the company's president, Joseph Gwinn, who wanted to determine exactly when the aircar's undercarriage would fail. While seated beside his test pilot, with slide rule in hand, Gwinn instructed Love to fully test the capabilities of his aircraft in what would be a terrifying experience for a pilot who'd been trained to land in a controlled glide. During the test, the aircar's shock absorbers performed so well that the two felt only a slight jolt. Although skeptical of its safety at first, Love tested Gwinn's aircar thoroughly and eventually she became convinced of its safety, always her chief concern as a pilot. Love had left the Gwinn Company by 1938, the year that aircar technology suffered what was to prove an insurmountable setback when a second aircar prototype struck high tension wires and crashed. The other test pilot and his passenger were killed.

Other Women

These four women — Gower, Raskova, Cochran and Love — were influential in bringing their counterparts into military aviation, but they alone could not have made such a significant contribution to the war effort without a pool of female aviators from which to form their women's units. The women pilots of World War II came of age during a time when girls were deeply influenced by the examples of famous women aviators. Several of these famous flyers wrote articles urging their contemporaries to learn to fly. Women throughout the world took flying lessons and joined flying clubs. Yet, it was the advent of the world war and their countries' prewar enhancements to military preparedness that provided the best opportunity for ordinary women to learn how to fly and greatly increased the number of potential female military pilots.

Preparations for Air War

The Treaty of Versailles, which concluded the First World War, prohibited Germany from manufacturing powered aircraft, but non-powered flight was allowed, so gliding clubs proliferated there during the interwar years. British youth maintained an amicable camaraderie with young Germans who flew in gliding clubs, even up to the weeks preceding the Second World War. Hanna Reitsch, the famous German glider pilot who later tested aircraft for the Nazis, visited England several times during the interwar period. Despite this friendly exchange based on a common interest, it was clear to anyone who observed closely in the latter years of the 1930s that preparations for war were well under way in Germany. Adolf Hitler, realizing the importance of air power, promoted "air-mindedness" among Germans who ranged in age from young children to university students. He also boosted the country's production of powered aircraft. British pilot Margaret Fairweather, while on an aerial honeymoon tour of Germany in 1938, expressed concern about the many *Luftwaffe* airfields she had noticed, proving Hitler's violation of the Treaty of Versailles.[13]

The threat of war brought a sense of crisis throughout the world. Many countries, concerned about Hitler's rapid rise to power, made plans to address shortages of equipment and personnel; but governments advanced those plans cautiously, attempting not only to mollify a nervous public, but also to appease potential opponents. The trick was to prepare for war without appearing to do so. In late 1934, Great Britain met the threat posed by German rearmament by forming the National League of Pilots, which would address the country's "air weakness." Just before the war started, Britain instituted the Civil Air Guard (CAG) and provided subsidies to flying clubs for training additional pilots. Even so, England entered the war with fewer pilots than airplanes, a fact particularly evident in the Battle of Britain. During those early years, every trained fighter pilot was critically needed for combat missions.

According to a speech by Joseph Stalin, the nonaggression pact that the Soviet Union concluded with Germany in 1939 gave the U.S.S.R. precious time during which to prepare for war.[14] Historians have since claimed that the pact allowed Stalin to let his country lapse into complacency before the surprise German invasion. In any case, at the outbreak of war in 1941, the Soviet Union had plenty of trained pilots, thanks to its campaign to develop competence in aviation; but the country seriously lacked adequate numbers of modern military aircraft. Women also learned to fly, encouraged by the Komsomol, a junior and subordinate branch of the Communist Party. The Komsomol sponsored the Soviet air force and offered flight training through paramilitary clubs. In the Soviet system, women had full constitutional equality with men, and though men were required to perform military service, women volunteers were not refused; but in actual fact, long-established tradition posed social, if not legal, barriers. For example, Soviet pilot Tat'yana Makarova, who later served in one of Raskova's regiments, the 46th Guards Regiment, trained at a flying club but was not allowed to enter a school for military pilots, while her less skilled male classmates were accepted.[15]

During most of the interwar period, the United States lagged behind other countries in air education, a fact that did not escape the notice of one well-traveled Canadian pilot who was making a prewar visit to her parents in Virginia. Helen Harrison was surprised to discover how much more air-minded other countries were during the 1930s. Harrison had traveled widely and she learned firsthand that other governments subsidized air clubs to make membership accessible, even to people of ordinary means.

The U.S. Army Air Corps' Henry Harley "Hap" Arnold knew aviation's potential in wartime and he promoted air power. Likewise, President Franklin Roosevelt was convinced that the country needed to meet the threat posed by the German preparation for war. However, even as late as 1939, the year that World War II began in Europe, the United States was ill-prepared for wartime aviation. Many Americans wished to maintain the country's neutrality, and many still remembered the First World War. Those who did perceive a threat from Germany in Europe and Japan in Asia also understood that a balance had to be struck between the need to prepare for war and the importance of maintaining an appearance of neutrality. The Civilian Pilot Training Program (CPTP) is an example of that kind of "under-the radar" initiative.

The CPTP began in 1939 with a two-fold purpose: economic recovery and war preparedness. As part of Roosevelt's New Deal, the program boosted the struggling American aviation industry, which had been hurt by the Great Depression; but it also generated many more pilots who could fly the ever increasing numbers of newly manufactured aircraft. The concept of civilian pilot training was proposed by Robert H. Hinckley, a New Deal administrator who decided that college classrooms and local flying schools could combine to offer the new government subsidized program. In July 1939, the goal of the U.S. Army Air Corps was to produce 1200 pilots a year, compared to the previous rate of approximately 200 annually. Hinckley used the term "air-conditioning" to describe his concept of cultural and technological adjustment to the air age. Although his idea paralleled the state-sponsored aviation training programs in Europe and England, the CPTP was controlled by the nonmilitary Civil Aeronautics Administration, unlike the model prevalent among the so-called "aggressor nations." By the end of 1939, four-hundred thirty-five colleges and universities in the continental United States and Alaska, Hawaii, and Puerto Rico were training new pilots.[16]

Women were accepted into the CPTP, their presence promoting an image of peaceful intentions while the program trained the large numbers of pilots that would be needed in wartime. Four women's colleges participated in the program — Lake Erie College (Painesville, Ohio), Adelphi College (Garden City, New York), Mills College (Oakland, California), and Florida State College at Tallahassee.[17] In women's colleges no predetermined ratio applied, but elsewhere women were accepted into the CPTP at the rate of one woman for every ten men. Although the country's many flying clubs already included women in their ranks, the number of women pilots increased substantially after the CPTP was implemented. Overall, the number of female pilots in America increased from 675 in 1939 to nearly 3000 in 1941.[18] Many personal accounts by American women who flew in the WAFS or WASP cite learning to fly through a CPTP program as the beginning of their path to military aviation.

After Japan's attack on Pearl Harbor drew the United States into the war in December, 1941, even the rather limited level of opportunity afforded through the CPTP was retracted for women because it was then clear that pilots in training would enter the military; and, while women were not required to enlist, the men were. First Lady Eleanor Roosevelt requested an explanation for the abrupt policy change, which even excluded women from being CPTP flight instructors, and Hinckley replied to her that the program for women would resume as soon as possible. Nevertheless, the preparations for war that were undertaken throughout the world became the catalyst that opened the door to military aviation for many women.

2

How Women's Units Came About

Great Britain: The Women's Section of ATA, 1939

Pauline Gower, the aviation entrepreneur, believed that women could play a role in wartime as aerial ambulance or ferry pilots, and possibly in other capacities too. Well known in England, thanks to her social standing as the daughter of a member of Parliament, not to mention the publicity her unusual career choice had generated, she was well positioned to use her social connections and considerable interpersonal skills, along with her prowess in aviation, to realize her plan for creating a women's section of the Air Transport Auxiliary (ATA). Gower wrote articles promoting the advancement of women in aviation. Although her writing shows that she considered aviation an important element in peaceful exchanges among nations, she became increasingly convinced that women pilots could play a role in the conflict if war should come. An accomplished woman in her own right, Gower served on several committees concerned with aviation and its regulation and she was in steady demand as a speaker. When war looked inevitable, Lady Loch, head of the Women's Legion Flying Section of Britain's Air Ministry, met with Gower to discuss ideas for employing women pilots. With her Air Trips partner, Dorothy Spicer, and another friend from Stag Lane, Amy Johnson, Gower assembled a group of licensed women pilots in the Women's Aero Club.

In early 1939, Gower attended a private conference to discuss how women could be employed in wartime aviation. Britain's Civil Air Guard already employed female flight instructors, so the concept of women as pilots was not unprecedented. Indeed, one of the commissioners of the CAG, Mrs. Maxine "Blossom" Miles, was selected on the insistence of Harold Balfour, parliamentary undersecretary of state for Air. On May 10, 1939, Gower became the Civil Air Guard's second woman commissioner in charge of the London and the southeastern areas.[1] After Miles resigned her commissioner's position, Gower was the obvious choice to lead a section of women pilots. Despite Balfour's support, there were many who considered training women for any work outside the domestic sphere to be a pointless exercise. Gower presented her plan to Lt. Col. Sir Francis Shelmerdine, the director-general of civil aviation, who approved of forming a women's section, speculating that it might consist of "perhaps twelve pilots" under her administration.² Having received Shelmerdine's positive response, Gower awaited further developments; but when time passed and she heard nothing, she inquired about the plan. Shelmerdine replied that it had been "shelved," citing reasons ranging from doubts about the female capacity for training to the lack of accommodations for women at air bases. Later, however, and with no explanation, in November 1939, Gower was directed to recruit a small number of women pilots for the ATA.

Britain's Air Transport Auxiliary had been formed in 1939 in response to a critical need for pilots to move new aircraft from factories to the airfields where they were fitted for combat. Gerard d'Erlanger, director of British Airways Ltd., which eventually merged with Imperial Airways to become the British Overseas Airways Corporation (BOAC), was its commander, with BOAC its administrator.³ At first ATA's staff consisted of twenty-seven Royal Air Force (RAF) pilots who flew Tiger Moths, single-engine open-cockpit airplanes that were used for training. As the war progressed, RAF pilots were considered far more valuable for combat than for ferrying airplanes. The answer to the need for ferrying pilots was to hire civilians who were too old for RAF service or whose physical limitations prevented them from joining the RAF. During its first year, both ATA and RAF pilots ferried aircraft, but by 1940 only ATA pilots performed the ferrying work. The ATA ultimately consisted of 1318 members, 166 of which were women. At its peak, ATA pilots and ground staff were stationed at twenty-two air bases scattered throughout Britain.⁴

The ATA organization was not itself considered experimental, but its Women's Section began with provisional status, pending satisfactory results. The men in the RAF, whose work the ATA supported, had to be convinced that women were up to the job of flying aircraft beyond light and simple trainers. Eventually those doubts were put to rest by the consistent performance of they, but first, they had to prove themselves.

Gower was appointed as a second officer in ATA and posted to its Hatfield Ferry Pool on December 1, 1939. Having passed her flight test in an RAF Tiger Moth, she was promoted to first officer and allowed to select other female pilots for the Women's Section.⁵ Gower gathered twelve of her peers — also referred to as "society women"— and flight tested them herself. Although she had wanted to choose ten pilots, she was instructed to limit her selections to eight. The eight women she assembled formed the nucleus of ATA's Women's Section and they began their work on January 1, 1940.⁶

Because of her extensive flying experience, along with her family's social connections, Rosemary Rees du Cros was among the first members of the Women's Section. Du Cros had the luxury of pursuing aviation as a hobby and had made several flights to Europe between the wars. The other seven women were Winnifred Crossley, Gabrielle Patterson, Joan Hughes, Marion Wilberforce, Margaret Fairweather, Mona Friedlander, and Margaret Cunnison. Four of them — Crossley, Hughes, Wilberforce, and Margot Gore, who joined shortly thereafter — remained with the ATA throughout the war.[7]

The first eight women became second officers. They were housed in private homes near the Hatfield Aerodrome. Early in their service the women were allowed to fly only light training aircraft not only because they weren't trusted in the cockpits of more advanced models, but also because "conversion" courses, which instructed pilots on more advanced aircraft, were unavailable to them because the RAF's Central Flying School, which offered these courses, was open only to men.

As the Women's Section grew, a second women's ferry pool (all female) was created at Hamble with Margot Gore as its commander. Gower established a club for the women pilots in London because there were no places for them to relax that were equivalent to the male pilots' service clubs, where women were not admitted. As the numbers of women in ATA increased, so did Gower's administrative responsibilities. To maintain contact with all her charges in the most efficient manner, she eventually relocated from Hatfield to White Waltham, the ATA headquarters.

Soviet Union: The 122nd Aviation Group, 1941

During the 1930s, Marina Raskova, an aviation record holder and the first female navigator in the Soviet Union, was as inspiring to young Soviet girls as Amelia Earhart was to their American counterparts. After setting the 1938 distance record in the *Rodina*, Raskova spent many months recovering from her ordeal in the wilderness. During that time she wrote her autobiography and diligently responded to fan mail. She had by then decided to pursue a military career and she studied at the M.V. Frunze Military Academy.

When the Soviet Great Patriotic War began, precipitated by the German attack in June 1941, Major Marina Raskova volunteered but was initially rejected for military service at the front; so she stayed in Moscow and became the moving force behind the formation of the women's air regiments. Joseph Stalin, the self-appointed leader of the Soviet Union, had taken great interest in the *Rodina*'s flight. Its crew — Raskova, Valentina Grizodubova, the pilot, and Polina Osipenko, the copilot—were received in Moscow with great ceremony after they were rescued and each was awarded the title Hero of the Soviet Union (HSU), the first women so honored.[8] Stalin had met all of them, and was acquainted with Raskova. Raskova knew that women were interested in serving their country as pilots because so many had written to her asking to be included in any air regiments she might form. As a member of the People's Defense Committee, Raskova made speeches urging her countrymen to join the battle against Germany. Large numbers of Soviets, including women, volunteered for military service. In fact, when the call went out, so many women responded that approximately one thousand could be selected for training. Raskova envisioned female air regiments serving alongside the volunteers from the Komsomol, the Young Communist League, who could train to be mechanics, staff personnel, gunners and navigators in support of those regiments. Despite her fame and charming personality, Raskova found that convincing the Soviet government to form aviation regiments consisting of women took several months of persistent effort. On October 8, 1941, she did obtain permission to assemble a flight group, and began interviewing candidates and choosing among the volunteers for military training. A good number of Soviet women were then members of aviation clubs, so many of Raskova's volunteers were already highly experienced aviators before joining her flight group.

Pennington claims that three factors led to the formation of the women's air regiments: a personnel shortage, Soviet propaganda, and the demands and persistence of Raskova herself. Pennington contends that there was no shortage of Soviet *pilots* in 1941, as commonly believed; rather, she cites evidence of a troublesome shortage of *aircraft*.[9] Cottam, who has translated the memoirs of many of the members of the women's regiments, concurs that the "shortage of pilots" myth enjoys wide circulation, but she thinks the myth should be retired. She notes an abundance of trained male pilots, a shortage of airplanes, and even a common societal attitude that women weren't suited to tasks like flying military aircraft.

Raskova personally interviewed all of the mostly civilian volunteers and she also directed their training. The three regiments would be distinguished by the type of aircraft they flew and the kind of flying they did—fighters, night bombers, and dive (day) bombers in the 586th, the 588th and the 587th regiments, respectively—and they were formed in that order. The 586th and the 587th regiments eventually included men in their ranks, mostly as ground support personnel, but the 588th Night Bomber Regiment would remain entirely female throughout the war. Raskova decided that she would command the last regiment that she had formed and trained, the 587th Dive Bomber

Regiment.[10] There were already many women serving in the Red Army, but the existence of women military pilots was unprecedented. For this reason, the number of potential female officers who happened to be familiar with aviation and therefore able to assist effectively was very low. Therefore, Raskova took almost full charge of the task of turning her civilian volunteers into a trained fighting force.

Americans in the ATA, 1942

In 1936, Jacqueline Cochran, by then a wealthy businesswoman and record-setting aviator, attended a party where she heard a German World War I flying ace, Ernst Udet, comment that Germany planned to go to war with America. Cochran brought her concerns about Udet's statement to the chief of the U.S. Army Air Corps, Major General Oscar Westover, but he brushed them off. Cochran was sure that the training provided to American pilots was inferior to that provided by Germany as Udet had described it. American pilots did not learn to fly with instruments, a skill Cochran had acquired by herself and one she considered essential in warfare. Besides, Udet had claimed that there were 100,000 military pilots in Germany, far exceeding the number of pilots in America. In 1939, on the day Poland surrendered to German troops, Cochran wrote to her friend Eleanor Roosevelt, recommending that women pilots be considered by the military for routine flying tasks as a way to free male pilots for combat.[11] Both the President and his wife favored employing women pilots in some way if America entered the war. Cochran's personal fame, and the business connections available to her via her wealthy husband, provided her unusual access to both Roosevelts, an advantage that helped advance her plans.

Cochran also presented her plan for creating a unit of women pilots to General Henry H. "Hap" Arnold of the U.S. Army Air Corps, hoping to gain his support too. Arnold was very skeptical, unsure that females could manage large multiengine aircraft. However, he approved Cochran's flight in a Hudson bomber from Montreal, Canada, to Prestwick, Scotland, a bid to promote the Lend-Lease program, which was ferrying airplanes across the Atlantic to England. To prepare for her flight, Cochran trained on a bomber, but her preparations met with resistance from male pilots whose exclusive right to fly that type of airplane she was challenging. They used all means at their disposal to impede her training, including delaying routine maintenance or removing essential tools from the plane on the runway. After debating with the rest of the crew over the extent of her role during the flight, Cochran and the pilots agreed to a compromise: she would neither take off nor land the plane, so in reality Cochran controlled the bomber only while it was airborne. Another pilot had the controls when it arrived in Prestwick on June 20, 1941.[12] One serendipitous benefit from her flight was that Cochran got a chance to spend ten days in England, and she used that time to meet Pauline Gower and to observe the workings of the ATA's Women's Section. This proved to be very useful preparation for Cochran when assembling her own women pilots in Montreal for service with the British, and later for her plan to incorporate women into the U.S. Army Air Force.

In *Global Mission*, Arnold, who commanded the United States Air Forces in World War II, describes meeting "Miss Jackie Cochran" in London and discussing the idea of "creating an organization of women pilots in the United States Air Force," although he does not cite the date of their meeting.[13] When Cochran returned from England after her bomber flight, she was hired by Lt. Colonel Robert Olds of the U.S. Army Air Force to be a "dollar-a-year" volunteer, tasked with developing a list of women pilots who could be recruited for service in England. Arnold had already given Olds permission to find fifty women who could perform domestic ferrying duties, but he had ordered Olds to delay implementing the plan. Cochran's concept for employing women pilots was broader than Olds' plan, which confined recruitment to no more than one hundred women, who would work only in the Ferrying Division. Cochran foresaw the need for much larger numbers of trained pilots in the event that America entered the war. She and her assistants discovered many likely candidates for an American group to fly with the ATA on the list of pilots who were registered by the Civil Aeronautics Administration.

Meanwhile, Arthur Harris, chief of the British Air Mission based in Washington, D.C., tried to increase American awareness of the work of the ATA and of England's huge need for pilots. President Roosevelt, having heard firsthand from Cochran about her impression of the Women's Section, knew that she wanted to form a similar group in America, but when he proposed that idea to Army Air Force personnel, he learned that there were no plans to put women pilots into military aircraft. Britain's critical need for pilots and America's need to determine whether women could fly military aircraft dovetailed in the summer of 1941, when Arnold asked Cochran to coordinate a plan with Harris to supply a number of women pilots to help allay Britain's pilot shortage *and* test Cochran's theory that women could indeed fly military aircraft.

On January 23, 1942, Cochran sent detailed two-page telegrams to women she'd chosen from the CAA lists, explaining the work they could do

Jacqueline Cochran in a publicity photograph with Captain Norman Edgar of Britain's Air Transport Auxiliary. The photograph was probably taken in 1942 as Cochran assembled women pilots to fly with the ATA. Their success made a similar program possible in America (courtesy Library of Congress).

as members of the ATA and inviting them to respond if they were interested.[14] One pilot, Dorothy Furey, had already anticipated that a group of women would be summoned to assist the British, so she made her way to Canada well ahead of Cochran's invitees. Furey and Cochran's personalities clashed, according to Whittell, who speculates that Cochran's negative response to Furey's independent action was typical of her behavior whenever the supremacy of her command was challenged.[15] Cochran interviewed her selections at various locations throughout the country, and those who met her test traveled to Montreal, where they were examined, inoculated for overseas travel, and flight checked by an ATA officer. Twenty-five of the women passed their flight checks and the initial physical exam. The twenty-four women, who also passed ATA's physical, made an early spring journey to England by whatever ships were available to transport them, a dangerous ocean crossing through seas patrolled by German U-boats. On October 26, 1942, First Lady Eleanor Roosevelt, who often served as the eyes and ears of her husband, visited the ferry pool at White Waltham. She showed particular interest in the Women's Section and the twenty-two American women serving there.[16]

Fahie notes that only one American woman proved unacceptable to the ATA and she was dismissed and sent home. Helen Richey returned to America in January 1943 after she ran a twin-engine Vickers Wellington into a hanger at full power on takeoff, an accident that seriously damaged the airplane but left Richey uninjured and able to walk away from the wreck. Safety was the top priority for the ATA because each airplane delivered was so desperately needed; the ATA consequently expelled Richey for damaging too many of them.[17]

United States: The Women's Auxiliary Ferrying Squadron (WAFS), 1942

As America hurriedly prepared for war, Nancy Love, who had worked as a test pilot and advocate for the Works Progress Administration's air-marking program, knew that America's entry into the war meant that additional pilots would be needed to transport new airplanes from the factories where they were manufactured to the ports where they would be shipped overseas. Love herself had flown American aircraft to Canada as part of the Lend-Lease program. Her husband, Colonel Robert Love, was deputy chief of staff for the Air Transport Command (ATC), where Nancy also worked, commuting by air from their home in Washington, D.C., to Baltimore, Maryland. Both

Loves envisioned a role in wartime for highly skilled female pilots, and some writers speculate that it was a casual comment by Bob Love about his wife's air commute one stormy morning that helped plant the idea of using women pilots into the minds of his superiors in the ATC.

Nancy Love wrote to Lt. Colonel Olds in May 1940, presenting him a list of highly experienced women pilots who could be called upon for ferrying work.[18] Olds was receptive to her proposal because his Army Air Corps Plans Division was interested in recruiting as many as one hundred women to ferry single engine aircraft. However, General Arnold did not yet want to implement the plan because he was sure there were sufficient male pilots to meet immediate needs and he was not yet convinced that women were capable of handling military aircraft. Meanwhile Lend-Lease deliveries of aircraft to England proceeded, with Cochran making her prewar bomber flight across the Atlantic to publicize the program. Months later Cochran gathered a small group of women pilots to assist the ATA.

Reconciling Two American Plans

While Love and Cochran pursued their separate bids to bring women into military aviation, the Army Air Corps itself underwent a major reorganization and expansion, and its name was changed to the U.S. Army Air Forces on June 20, 1941.[19] Colonel Olds, of the Air Transport Command, whom Nancy Love had originally contacted, was replaced in April 1942 by Colonel (later General) Harold L. George. It was George who gave Love permission to proceed with her proposal for women pilots, but he did so without waiting for authorization from his own commander, General Arnold. Therefore on September 5, 1942, Nancy Love contacted eighty-three highly experienced women pilots, offering them a chance to interview and be flight checked for her new Women's Auxiliary Ferrying Squadron.[20]

The *New York Times* featured an article about the formation of the WAFS, including a photo of Love, Secretary of War Stinson, and General George on September 11, 1942 — just in time for Cochran to see it upon her return from her six months' assignment with the ATA.[21] Whether General Arnold had approved the formation of the WAFS, or whether his approval had been assumed to the extent that Love's plan was actually implemented without his permission, remains unclear. In any case, for a short time there were *two* plans to accept women pilots into the United States military. Differences of opinion about the origins of these American women's units and the roles Love and Cochran played in their formation persist to this day.

Cochran was outraged to read about Love's new group of American women pilots. She had had General Arnold's promise that her own training program would commence as soon as her women pilots' ATA performance was judged acceptable. Knowing they had performed well alongside their British counterparts and furious at being circumvented, Cochran protested to General Arnold, who angrily claimed that he was unaware that Harold George had authorized the creation of a women's squadron in the Air Transport Command. Cochran was also sure that recruiting so few pilots as the number anticipated for the WAFS would not address the need.

The problem of reconciling Love's existing ferrying squadron with Cochran's proposed pilot training program was addressed by General Arnold when he placed Love in charge of the women in the Air Transport Command, but with no future plans for increasing their numbers. The small number of WAFS would function within a much larger organization of women pilots that would be commanded by Cochran. Therefore, Cochran's group came into being almost simultaneously with the WAFS, but Love's pilots are appropriately, and accurately, called "the originals" because they were the first women ever authorized to fly military aircraft for the United States.

United States: The Women's Flying Training Detachment (WFTD), 1942

On September 15, 1942, General Arnold announced that Jacqueline Cochran would head a training program for licensed women pilots who would be instructed in all kinds of flying *except* combat. Cochran chose the Howard Hughes Air Base near Houston, Texas, for her training program's home. In November 1942, she established the Women's Flying Training Detachment (WFTD), officially designated the 319th Army Air Forces Flying Training Detachment.[22] Classes in the WFTD were designated by the year of their graduation and the sequence of their entry into the program. The first class, 43-W-1 ("W," of course, stood for "women"), reported for training in Houston with new classes arriving every thirty days. The rapidly expanding flight school soon outgrew its original location, so the forty-five women in the second WFTD class, 43-W-2, flew to Avenger Field in Sweetwater, Texas, an airfield that had previously served as a training base for male pilots and would now be the new home of Cochran's trainees. Cochran was not pleased when she realized that all of the graduates of WFTD class 43-W-2 had been assigned to air bases selected by Love and would therefore be working in the Ferrying

General Henry Harley "Hap" Arnold at his desk in the Munitions Building in the 1940s. At first doubtful that women could fly large military aircraft, Arnold later became a strong advocate for the WASP (courtesy Library of Congress).

Command, so she lobbied for greater control over the program to insure that the women were trained to perform all kinds of military flying, not just cross-country ferrying.

United States: The Women Airforce Service Pilots (WASP), 1943

The WAFS and the WFTD merged into a single unit on August 5, 1943, becoming the organization known as the Women Airforce Service Pilots.[23] General Arnold appointed Cochran as director of women pilots, and Love as executive of the Ferrying Division of the Air Transport Command, and, thus, subordinate to the leader of the WASP.

In their attempts to bring American women into military aviation, Love and Cochran proposed two very different plans. Love wanted to recruit a small number of highly experienced pilots who would require little training and whose chief duty would be ferrying. The fact that the recruitment of those particular pilots proceeded so slowly — their numbers never approached the figure that Love expected to achieve — perhaps indicates her plan's intrinsic limitation. Aviation, with few exceptions, was the realm of the well-to-do. It was beyond the scope of women of modest means and few were willing to sacrifice to pay for expensive hours aloft. There were limited opportunities to participate in quota-driven programs like the CPTP, so the number of women pilots who could meet Love's stringent requirements was relatively low. Cochran wanted to recruit large numbers of women who had logged fewer flight hours but who could be taught a wider variety of flying, in addition to cross-country. Love's highly experienced squadron satisfied existing ferrying needs, but Cochran's plan to train many more pilots was designed to meet anticipated needs. The competition that developed between members of the WAFS and the WFTD appears to be grounded in the contrast between these two visions and in the tensions that already existed between the Air Transport Command and the Army Air Forces. The merger into the WASP brought clarity to the scene, if not harmony, and it established a chain of command for the women's units. Cochran's plan prevailed, partly because she could argue effectively; but the fact that Cochran and Arnold had developed her training plan together must have played no small part. Cochran's role eventually became more firmly established when she was named director of all women pilots flying for the Army Air Force and assistant to the chief of Army Air Forces staff in Washington, D.C., in June 1944.[24]

3

Women's Flight Units

The threat of war's arrival precipitated massive preparations throughout the world, including the first serious consideration of using women as military pilots. England's declaration of war on Germany, Germany's invasion of the Soviet Union, and Japan's attack on Pearl Harbor in the United States brought all three countries together as allies. The women who were already pilots sought to use their knowledge of aviation in their country's war efforts.

Motivations

The women in established aviation units were the first women actually *recruited* to fly for their countries' air forces. They joined for different reasons, based on their own life experiences and their country's circumstances in wartime. Their passion for flight manifested itself early in life and most of them loved the physical sensation of flying and were enchanted with the sense of freedom they felt in the air. A few women, on both sides of the Atlantic, write in their memoirs that they were thrilled during childhood by the story of Peter Pan, the British tale animated by Walt Disney in 1942.[1] In postwar accounts, nearly all the women tell of their passion for flying, so it seems that their primary motivation for becoming military pilots was having a chance to work at something they loved. Rosemary Rees du Cros writes in her memoir that although pilots would misbehave while off-duty, no one in the ATA risked missing a flight for fear of losing that much cherished occupation. She notes that ATA pilots could be "sacked" for failing to meet a flying obligation.[2]

Veronica Volkersz was placed on a waiting list after her flight test at Hatfield in 1940, having logged fewer flight hours than the other women tested that day. It wasn't until March 1941 that she was accepted. Meanwhile, Volkersz drove an ambulance for the Women's Auxiliary Air Force (WAAF) but she writes that while she waited, she never passively accepted her "waiting list" status but "hammered away" until she was allowed entry.[3] Some members of the British WAAF joined the ATA in 1943 when the Royal Air Force allowed thirty women (of the 1400 who had volunteered) to take an ATA flight test.[4] Escott, in her history of the WAAF, writes that the response to the RAF's call was "overwhelming." The women selected were granted a temporary release from their WAAF duties. One of them, Peggy Eveleigh, says her "first day of training was one of the most exciting of my life — only beaten when I flew my first Spitfire in September 1944."[5]

The average woman's access to flight training in peacetime was limited, but the contrast in opportunity for female and male flight students became even more striking in wartime. Flight training was provided at no cost for men accepted into service, but women were denied this benefit because they were not expected to serve as wartime pilots. The British and American women trainees were the only female pilots in their countries who enjoyed the benefit of military flight training. Women who joined government-subsidized flight training programs during the prewar years often did so to help defray the cost of flight lessons. Volkersz writes that she joined England's Civil Air Guard as a way to help her pay for flying lessons.[6] American Elizabeth Strohfus, 44-W-1, recalls using her brand new bicycle as collateral for the $100 loan she arranged to join a local sky club. Strohfus joined the Civil Air Patrol (CAP), but discovered that she was the only "girl" in the group. She writes in her memoir that she felt "selfish" about having such a great time in the CAP, so she invited other women to join, eventually adding five women to the group's thirty men.[7]

Soviet women were encouraged by their government to pursue aviation and they took flight lessons in the *Osoaviakhim*, the Society for the Defense, Aviation and Chemical Industries.[8] Many women were already working as flight instructors when Germany attacked in June of 1941. Thousands of them rushed to military recruiters, but at first they were turned away. They also sent letters to Marina Raskova, begging her to get them to the front in a Red Army Air Force unit.[9] Senior Lieutenant Sasha Krivonogova, a pilot in the 587th Day Bomber Aviation Regiment, whose daughter had become ill and died during an evacuation, is just one example of why the reasons for volunteering could often be intensely personal.[10] Grizodubova's recollection about one of the pilots explains her motivation: "The Germans killed her mother, father, and one of her children — only her younger son survived. Her hatred for our enemies was as great as her tenderness for her surviving child. None of her bombs missed their aim." Grizodubova describes another pilot who'd lost her entire family: "In her dark eyes lay the shadow of profound grief, but her handshake was as firm as a man's. To avenge the dead, to assure a better future for the living — that is why we Russian women went up to battle the Nazi foe."[11]

American pilot O. Vivian Hicks Fagan, WASP class 44-W-7, who had worked in the CPTP in Lind, Washington, withdrew her application to the Women's Army Auxiliary Corps (WAAC), in favor of applying to the WASP, after hearing her friends describe their experiences. Fagan was accepted, and filled the five months she had to wait before her training began by working as a flight instructor in Pullman, Washington. She considered instructing the best way to "give back" the knowledge the United States government had provided her through the CPTP.[12]

Tennessee native Cornelia Fort was airborne over Hawaii in December 1941, helping her flight student practice takeoffs and landings, when the Japanese attacked Pearl Harbor. When Fort realized that the airplanes surrounding her were neither American nor friendly, she hurriedly seized the controls, landed the plane, and dashed to a nearby hanger as Japanese fighters strafed the field. The attack resulted in an immediate cessation of all nonmilitary flying, preventing Fort from joining Cochran's ATA group. Grounded in Hawaii, she wasn't able to reach the Canadian departure point on time. As soon as she could book passage by ship, she crossed the Pacific and joined Love's WAFS, who were then gathering at the New Castle Army Air Force Base in Wilmington, Delaware.[13]

Evelyn Sharp was a highly experienced young barnstormer from Nebraska, but she declined Cochran's offer to join a group of women pilots bound for England. In *Sharpie: The Life Story of Evelyn Sharp, Nebraska's Aviatrix*, Bartels speculates that Sharp anticipated the formation of a unit for women pilots in the United States and decided to wait for that opportunity instead.[14] Her choice to gamble on an American group was vindicated when she received another telegram in September 1942 from Harold L. George, the new brigadier general of the U.S. Army's Air Transport Command, inviting experienced women pilots to join the WAFS.[15]

Requirements and Eligibility

Many of the pilots who began their aviation careers in the barnstorming and joyriding era had aged beyond eligibility for regular military service by the time the Second World War began. Eligibility requirements for the women's aviation units varied from country to country and, over the course of the war, in individual countries, so there wasn't one consistent standard for acceptance, except perhaps the most basic of requirements: candidates had to be young and physically fit, although even those standards were relaxed over time. The urgency of their need for pilots to ferry aircraft or perform other routine tasks also determined the extent to which the rules for admission were stretched in Great Britain and the United States. In the Soviet Union, a Party committee made initial acceptance decisions, but afterwards, Marina Raskova interviewed each applicant who had cleared that hurdle. Both American and British women submitted to physicals before being accepted. There was no physical examination required for Soviet women pilots.[16]

Standards for admission into the ATA in Great Britain were rigorous early in the war, particularly for the women, so there weren't many female pilots in that country who qualified. The requirement of 500 logged solo flight hours would have been difficult even for women of means to attain.[17] The acceptable age range was between twenty-two and forty-five.[18] However, by mid–1942, requirements for the Women's Section were reduced to 50 hours' flight time and the possession of an "A" license (which did not allow passenger transport).[19] Later, even those requirements were reduced. Diana Barnato Walker worked with the British Red Cross, first as a nurse and later as an ambulance driver. During that time, Walker met Lois Butler, an early member of the Women's Section. Walker, intrigued by the possibility of flying with the ATA, took its flight test after she had logged only *ten* solo flying hours, which had been recorded two *years* earlier. In her memoir, *Spreading My Wings: One of Britain's Top Woman Pilots Tells Her Remarkable Story*, Walker acknowledges that Britain, during the period following the Battle of Britain and before America entered the war, was "scraping the bottom of the barrel."[20] In early 1943, the shortage of pilots in Britain was so acute that women were allowed to join the ATA and then were trained to fly. Monique Agazarian, having never before flown *in* an airplane, applied to the ATA and was accepted as one of the ten women recruited for "*ab initio*" training.[21] According to Lettice Curtis, the height requirement for ATA women was at least 5 feet, 5 inches.[22]

In the Soviet Union, pilots for the new women's aviation group were chosen by the Komsomol, the name a syllabic abbreviation of the name Kommunistichesky Soyuz Molodyozhi, the Communist Union of Youth.[23] Candidates who successfully navigated the Komsomol's review process were then interviewed by Marina Raskova, who evaluated their qualifications and determined which regiment they were suited for and what roles they would play. Raskova alone determined each woman's eligibility and she assigned them to regiments on the basis of need. The most experienced pilots went to the fighter regiment, but all the regiments required navigators, mechanics, and other ground crew workers. Therefore, the less experienced pilots were diverted into training programs for navigators, while those with greater physical strength were tapped to work as mechanics and armorers.[24] No physical examination was administered, and one woman pilot notes in her memoir that during her time with the regiment she never saw a physician.[25]

In October 1942, twenty-eight American women were chosen for the WAFS, from among those meeting the following requirements:

1. Age 21–35
2. Height [minimun] 5′ 2″
3. HS [education] or equivalent
4. 200 HP rating (50 hrs. last 6 months)
5. 500 pilot hrs.[26]

The extent of this first group's experience is impressive. The originals had an *average* of 1100 flight hours.[27]

The first WFTD class consisted of twenty-five women, each holding a commercial pilot's license and meeting these requirements:

1. Age 21–35 inclusive
2. Height [minumum] 5′ 2″
3. HS [education] or equivalent
4. 200 pilot hrs.[28]

Frances Roulstone, 44-W-4, took flying lessons so she could qualify for the WASP when the minimum requirement for flight hours was reduced to thirty-five.[29]

Who They Were and Where They Came From

There was no typical female pilot during the Second World War. The women who volunteered for service — and all were volunteers — varied widely in background and personality.

The Women's Section of the Air Transport Auxiliary has the distinction of being the most international of all the women's aviation units. The ATA's total membership, both men and women, represented twenty-eight different nationalities.[30] The prewar British Empire covered much of the globe, and it was the scope of Britain's hegemony, combined with the fact that women were restricted or forbidden entirely from serving as pilots in Canada and Australia, that made the Women's Section of ATA the most international of all the women's units.

The Royal Canadian Air Force did not admit women despite a severe shortage of pilots. No one kept statistics on how many women applied for work as pilots in Canada, because the idea of women pilots in the RCAF was never entertained. Even ferrying tasks were unavailable to them. By January 1944, when the possibility of employing women pilots was considered, no women's units were actually formed. Margaret Littlewood was Canada's only female Link instructor, but that role kept her grounded in a pseudo-airplane, the Link trainer, a device used for teaching pilots "blind" (instrument) flying. Littlewood was hired by the manager of No. 2 AOS (air observer school) in Edmonton, Wop May, but her position was the full extent of those that were open to women pilots in Canada, where women in the military were expected to fulfill the motto "We Serve That Men May Fly."[31] Canadian women pilots joined the American WASP and others flew in the Women's Section of ATA. Helen Harrison was the first Canadian woman in ATA. She was followed by Gloria Large, Violet Milstead, Marion Orr, and Elspeth Russell Burnett.[32] Canadian Virgina Lee Warren was a WASP posted to the Romulus (Michigan) Ferry Command Base.[33]

According to Thompson's *The WAAAF in Wartime Australia*, Australian women pilots who might have joined the ATA were not permitted to leave the country.[34] Australia's shortage of airplanes, combined with large numbers of eligible male pilots, worked against the creation of a women's program there. Women performed a wide variety of roles in the Women's Auxiliary Australian Air Force (WAAAF), including flight mechanic, flight rigger, and Link instructor. Of the 27,258 women who served, none were employed as pilots.[35] Their work can best be described by the motto on a typical recruiting poster: "THEY [the women] KEEP THEM [the men] FLYING."[36] Australian women tested parachutes with logs attached, wooden "dummies" that they dropped from airplanes piloted by men. One woman in the WAAAF Transport Section, drove, but she did not fly. She recalls testing a parachute by dropping the "A.C.1 Wood" from her position behind the pilot's seat in an airplane.[37] Nicknamed "Airscrew Alma" for walking around so often with a propeller (airscrew) under her arm, one woman was a flight mechanic in Point Cook. Another worked on aircraft engines at the No. 6 SFTS (Service Flying Training School) in Mallala in South Australia. She writes, "After each inspection, a member of a section, mechanic, rigger, electrician, etc., would have to go on a test flight. That was a bonus as we loved flying."[38]

Both societal and official forces prevented qualified Australian women from flying military aircraft. Nancy Bird, an accomplished commercial pilot, writes in her memoir that Australian defense minister H.V.C. Thorby was convinced of the biological unsuitability of women to the physical demands of flight, yet he did recognize the existence of exceptions, like New Zealand's Jean Batten or England's Amy Johnson. During the war years, Bird served as the second commandant in the Women's Air Training Corps (WATC), which formed the nucleus of the WAAAF. Her task was to survey women pilots in Australia, with the goal of recruiting them for the ATA, but that plan never came to fruition because of the restriction forbidding women to leave Australia. Therefore, only Australians Mardi Gething, already in England when war broke out, and Victoria Cholmondeley actually served in the ATA.[39]

Several sources cite at least two women employed as pilots in South Africa in the Women's Auxiliary Air Force, but they are usually unnamed. Dolores Theresa "Jackie" Sorour Moggridge was no doubt one of them. Moggridge had learned to fly in her home country and moved to England in June 1938, on a quest for her "B" license (allowing passenger transport).[40] Her mother unsuccessfully tried to convince the young woman to return to Africa if war broke out. Lacking the requisite number of flying hours she'd need for immediate acceptance into the ATA, Moggridge drove an ambulance in the early months of the war. When the war's scope broadened and more pilots were needed, she was summoned to Hatfield Aerodrome, where she was flight tested by one of the eight original members of the Women's Section.[41] Another South African woman who joined the ATA, Mrs. Rosamund Everard-Steenkamp, had flown for the South African Women's Volunteer Air Force. Everard-Steenkamp was killed on one of the last of ATA's flights during the war.[42] It is possible that another of the South African female pilots was the Canadian-born Helen Harrison, the first woman flight instructor in South Africa.[43]

Women in ATA also hailed from New Zealand, Poland, Chile, Canada, Ireland, Holland, and the United States, according to Lettice Curtis.[44] Whittel adds Argentina to that list.[45] Three Polish women, Anna Leska, Stefania "Barbara" Wojtulanis, and Jadwiga Pilsudska, made their way to England after Poland was invaded by Hitler's army.[46] Margot Duhalde, a nineteen-year-old Chilean who held a commercial license, crossed the Atlantic on the *Rangitata*, a Norwegian-registered freighter that landed in Liverpool. Duhalde had hoped to fly with the Free French because of her French heritage; but that possibility did not materialize, not only because women never actually flew in France's air force, but also because France capitulated to Hitler so early in the war. By several strokes of luck, Duhalde, who spoke no English and never set out to fly airplanes in England, managed to connect with Pauline Gower, who was skeptical of the Chilean's ability to function without at least a rudimentary knowledge of the language. Gower was persuaded to give the potential recruit three months of training. Duhalde ultimately became a first officer in the Women's Section of ATA and became known by the nickname "Chile."[47]

Americans joined the ATA for an eighteen-month term of service. One of them, Hazel Raines, was already known as "Macon's First Lady of the Air," thanks to the *Macon (Georgia) Telegraph*.[48] Raines was the first woman in Georgia to earn a commercial license.[49] The letters she sent home during that time hint at the hazards associated with ocean travel in wartime. Roberta Sandoz Leveaux writes that she was one of *seventy-six* recipients of Cochran's telegrammed invitation to flight test at Dorval Airport near Montreal.[50] How many telegrams Cochran actually sent is not entirely certain. Whittell writes that she contacted one hundred twenty-five women.[51]

While they waited in Canada, Leveaux and the others were inoculated before leaving the country, and then flight tested. ATA Officer Harry Woods tested Leveaux just after the crash on takeoff of a heavily loaded B-26 bomber. She passed the accident-abbreviated check-ride, unlike several other women who had failed and had already returned to the U.S. Twenty-five women, among them two Canadians, one Dutch, and twenty-two Americans, were accepted by the ATA and they prepared to cross the Atlantic by whatever transport became available whenever departure was possible. Leveaux travelled with Emily Chapin, Myrtle Allen, and Mary Nicholson on a Norwegian freighter, the *Mosdale*, a ship considered agile enough to cross the ocean unescorted. Real dangers accompanied the Atlantic crossing and most ships sailed in convoys for protection. However, in an emergency, other ships in a convoy could not stop to assist because doing so made them potential German U-boat targets. During the crossing, the captain sometimes took evasive action based on information unknown to its nervous passengers. Leveaux and her shipboard companions landed in Liverpool. The women were fitted for uniforms in London and sent to White Waltham for ground school.[52]

All of Cochran's choices to augment the ranks of the ATA crossed the Atlantic safely. Pauline Gower met one group at its landing site in Liverpool, but the new arrivals failed to comprehend that Gower's invitation to them to dine with her that night was obligatory, not optional. The women chose to rest after their long voyage rather than join their new commander at dinner, inadvertently making an unfortunate first impression. This episode illustrates that the Women's Section of ATA was already operating under clear organizational protocols within a military context, a way of life still unfamiliar to the newcomers. This gaffe aside, Cochran's pilots were generally well received and successful in their work.

The earliest members of the Women's Section occupied the same position in British society as Gower herself, that is, they were part of upper class British society and were generally well off. The South African Jackie Moggridge, a person with more modest origins, remembers her discomfort when she first joined Gower's group. Cochran, the self-confident and brash American, did not fit harmoniously into their social context either. Her reception was not enhanced by the fact that her role in England seemed ambiguous. The ATA pilots may have resented Cochran's failure to do any flying at all while she was there, despite the critical need for every pilot who could possibly help. Even Gower accepted flying tasks whenever her administrative duties would allow. Photographs from that period depict Cochran, who held the honorary title "flight captain," standing in full ATA uniform along with other members of the Women's Section; but she considered her role in ATA to be a coordinator and observer rather than a participant.[53] It was Cochran's carefully selected group of pilots who would actually ferry the aircraft in England.

Histories of the Air Transport Auxiliary make no mention of its racial composition, but common societal attitudes are obvious in this excerpt from Gower's *Women with Wings*: "From early morning until late evening I seemed to be in the air and Dorothy had to work like a nigger keeping the air-frame and engine up to scratch."[54] Although it drew women from throughout the world, the Women's Section of ATA was not racially diverse. Gower sheds additional light on the era's racial context in *Women with Wings*, as she describes overhearing one couple's conversation about the impending war. They perceived that Japan would be no real threat in aviation, because they were sure that the Japanese were incapable of making accurate judgments of speed and distance. Gower criticizes the couple for speculating about the potential outcome of war rather than seeking ways to work for peace, suggesting that aeroplanes might be used to increase friendly feelings among nations.[55]

Soviet women's regiments were hastily assembled, using available records on current aviators and drawing from the hundreds of women who begged Raskova for a chance to serve as pilots. The compelling crisis brought by the German invasion was one reason that most of the women pilots in Raskova's regiments were ethnic Russians who were from Moscow, her home base, or its environs.

The Heroic Flight of the "Rodina" includes statements implying that the Soviet government's official policies toward women guaranteed them complete equality. Grizodubova, the *Rodina*'s pilot, says, "Soviet women are performing heroic feats of Socialist labour in all spheres of human activity."[56] In light of that claim, it is no surprise that Soviet women constituted a large part of the workforce and that they pursued more nontraditional roles in their own country than women elsewhere. Aspiring pilots were identified through air clubs, universities and technical schools, and factories. They were young, some only in their teens, and they were mostly single, although some of the women were married and a few had children. Women who were already serving in the Red Army became Raskova's best source for potential leaders of the new regiments. The volunteers' level of education varied, as did their experience and skill.[57]

Prevailing attitudes toward race in the United States prevented African-American women from serving their country as pilots in the 1940s. The picture of American aviation between the world wars mirrored the country's segregationist policies. African American women pilots certainly existed, but none were members of the Ninety-Nines, the organization of women pilots formed by Amelia Earhart in 1929. Black pilots had little choice but to follow separate tracks from their white counterparts. A few black women applied to the WASP, but Cochran rejected their applications, citing the tenuousness of her experimental organization. However, according to Caroline M. Fannin in her article "Aviation" in the encyclopedia *Black Women in America*, women of color were accepted into the Women's Army Auxiliary Corps (WAAC) and the ratio of black to white women was 80 to 440 in the first officer-candidate class for the Air WAAC. Black women pilots also served in the Civil Air Patrol, performing premilitary training and air-sea observation and rescue. Fannin also mentions Willa Brown, who served in the CAP along with American-born Earsly Taylor Barnett, who held Jamaica's first commercial pilot's license.

First Lady Eleanor Roosevelt championed the causes of both female and black pilots. She visited the flight school in Tuskegee, Alabama, touring the facility by air with chief Charles Alfred Anderson.[58] As the wife of the president, Mrs. Roosevelt was well-placed to promote the cause of equal opportunity, but there were substantial barriers. In *Black Wings: The American Black in Aviation*, black male pilots appear throughout, but black women appear only in photographs from the interwar years. An instructor and a woman trainee are pictured at the Coffee School of Aeronautics in Chicago, Illinois.[59] Images of black women pilots do not exist in the section about the war years. Black males were trained at Tuskegee AAF from 1941 to 1945 to see if black men could become pilots, giving them a certain equivalency with the women in the WASP program, also deemed experimental. However the Tuskegee Airmen were allowed to fly in combat, thus proving themselves in battle and ultimately resulting in the July 26, 1948, executive order of President Harry Truman, which ended the United States military's official policy of segregation and opened military career advancement to black men.[60]

In *The Stars at Noon*, Cochran writes that several "Negro" girls applied to her training program, and although she claims to have interviewed them "without preference or prejudice," she also admits that she wouldn't have known what she'd have done if any had passed all the preliminary tests. Cochran invited one of the women, an unnamed New Jersey schoolteacher, to Sunday breakfast in her New York City apartment, making a special trip east for that purpose. In her book, Cochran tells about explaining to a "Negro" applicant why her training program couldn't admit women who were black. When Cochran listed the practical difficulties that would ensue if the applicant persisted, she withdrew. Cochran writes, "This fine young Negro girl recognized the force and honesty of my arguments, stated that first of all the women pilots' program should be stabilized and strengthened, and she withdrew her application. She also saw to it, I believe, that I was left alone thereafter so far as this particular issue was concerned. I appreciated her understanding and respected her as a person."[61] There are other sources that cite Cochran's concern that southern Jim Crow laws would prevent black pilots from finding accommodations there. Cochran was also worried that her flight training program would not survive if it accepted black women.

A letter from Jacqueline Cochran, dated 19 August 1943, responds to an inquiry on August 4 from Sadie Lee Johnson in Tunica, Mississippi. Cochran writes that "there is no provision for the training of colored girls in the Women's Flying Training program" and suggests that the applicant investigate the WAC, who do "enlist colored girls for various types of work," or look for opportunities to serve the country by working in her own hometown.[62] Mildred Hemmons Carter received a similar letter when she inquired about joining the WASP, after having trained at the Tuskegee Institute's Civilian Air Training School. Janet Harmon Waterford Bragg also wanted to join the WASP, but was rejected by her interviewer who told her that—because of the Jim Crow laws—there would be no place for her to stay in Sweetwater, Texas. In her autobiography, Bragg describes her interview with a startled Ethel Sheehy, the chief recruiting officer for the WFTD, who exclaimed, "I've never interviewed a colored girl" and reminded her that WASP training occurred in a *southern* town. Sheehy forwarded Bragg's application to Cochran, but Bragg was informed, "Whatever Mrs. Sheehy told you (in the interview) still stands."[63] Rose Rolls Cousins, the first black female solo pilot in the CPTP, had a similar experience. Dorothy Lane McIntyre, the first black female in the United States to earn her private pilot's license in 1940, was also rejected by the WASP because of her race.[64]

WAFS Adela Scharr's *Sisters in the Sky* illustrates the racial context in America during the war years. First, Scharr was taken for a black woman by several black enlisted men, at a distance, who were disappointed to discover that she was actually white. Scharr resolved never again to get so tan![65] In the second instance, while hospitalized during her service, Scharr learned one day that there were guards assigned to a black male patient because the man's doctor was convinced of the "emotional instability of negroes."[66]

According to Haynsworth and Toomey in *Amelia Earhart's Daughters: The Wild and Glorious Story of American Women Aviators from World War II to the Dawn of the Space Age*, members of two groups — black male pilots and white female pilots who were training on the B-25 Mitchell bomber at Mather Air Force Base in Sacramento, California — enjoyed a mutual respect, each having felt the sting of discrimination. Both were in units that were still considered provisional by the Army Air Force. The black men and white women shared a mess hall at the air base, but the nineteen women were always served first. When the women noticed that they were consistently given priority, they asked the staff to end their special treatment.[67]

There were only two minority members in the WASP, women accepted into classes on both ends of the organization's short life span. Hazel "Ah Ying" Lee, 43-W-4, had flown commercially and privately in China during the 1930s and she held dual Chinese and American citizenship. In 1938, while she was living in China, Lee was displaced by the Japanese attack and she returned to America, helping to procure war materials for her country while residing in New York. Lee decided to join the WASP so she could participate in America's war effort, and eventually she was one of the one hundred thirty-two American women trained to fly pursuit (fighter) airplanes. Lee was stationed at Romulus Army Air Base in Romulus, Michigan.[68] During one delivery flight, she made an emergency landing in a field that looked acceptable, but when she climbed out of the airplane, she was pursued by a pitchfork-wielding farmer who was terrified of Japanese invasion and unaware that American women were transporting airplanes within the United States. Lee managed to convince him that she was not Japanese, but Chinese, and furthermore that she was on an airplane delivery mission for the United States. In time, the farmer relented and he even telephoned the Sweetwater airbase on her behalf. Lee introduced her WASP friends to Chinese culture, and assisted them with their orders in Chinese restaurants. One of her fellow WASPs states that having a minority friend was a unique experience.[69]

Maggie Gee, 44-W-9, born in Berkeley, California, in 1921, was the second Chinese-American woman pilot accepted into the WASP. Gee had learned to fly in Minden, Nevada, and she was a student at the University of California at Berkeley when Pearl Harbor was attacked. A few months after she graduated from WASP training, Gee was assigned to Nellis AFB in Las Vegas, Nevada. Because she had entered the training program so late, Gee served in the WASP for only one year, ferrying aircraft and towing targets for gunnery training. After WASP deactivation, Gee joined the U.S. Air Force Reserves, but she was nearly inactive and served only briefly. She later pursued a career as a physicist. Gee claims that her WASP experience gave her self-confidence. Although her associates in the program had never encountered a Chinese person before, Gee writes that she never had a bad experience.[70]

Status in the Military

Awarding military status to women who flew for the three Allied air forces was not a consistent practice. Their status ranged from absolutely civilian, with no expectation of becoming military personnel, to having constitutionally guaranteed equality as full members of their country's air force.

Militarization was never an option for Britain's Air Transport Auxiliary — it was founded to assist the Royal Air Force as a civilian organization and remained so throughout its existence. Pilots in the Women's Section had no expectation of being anything other than civilian pilots who supported the RAF. Under the management of BOAC there was no expectation among ATA's members that they were — or ever would be — part of the military.

In October 1941, the Soviet People's Commissariat of Defense ordered the creation of three aviation regiments composed entirely of women who would be trained to fly in combat. The women in Raskova's Aviation Group #122 took a military oath at the Engels training base on the anniversary of the Russian Revolution, November 7, 1941.[71] The three regiments were largely, or entirely, female for the duration of the war. Some of the women from Raskova's regiments were sent to reinforce men's regiments as needed. For the women who were already serving in the Red Army, it was their proximity to the battlefront that determined whether they chose to remain with their units or join Raskova's. The desirable choice, in their eyes, was the location closest to the battlefield.

When the women's units were formed in the United States, they were considered civilian, but their status was somewhat ambiguous for their entire existence. Most of the women, as well as their two leaders, anticipated a trend toward militarization, despite the fact that they began their service as civilians. Colonel William Tunner, commanding officer of the Ferrying Division, proposed militarizing the WAFS as soon as the squadron was formed. However, the time required to amend an already pending Women's Army Auxiliary Corps militarization bill was considered an unacceptable delay in activating the group of ferry pilots, so the WAFS entered as civilians and remained so.[72] The decision to forego the legislative process, in favor of early activation, had far-reaching and unanticipated consequences.

Jacqueline Cochran's long term plan for her pilots was that they would become officers in the Army Air Forces. She would not accept the proposal that placed them into the WAAC, then pursuing its own militarization, because doing so required her pilots — and incidentally herself— to be placed under the command of Colonel Oveta Culp Hobby, whom Cochran heartily disliked for Hobby's admitted ignorance of aviation and because Cochran considered her management practices slipshod. When General Arnold raised the possibility of placing her organization into the new Women's Army Corps (WAC), Cochran rejected his recommendation by turning the tables, saying, "How would *you* like to have the Air Force back in the Signal Corps?"[73] In fact, General Arnold had no way to order Cochran, a civilian, to comply with his proposal. The Women's Auxiliary Army Corps became the Women's Army Corps on July 1, 1943.[74]

Having civilian status meant that American women pilots lacked government insurance coverage. Strohfus writes that the WASP flew with no insurance at all — private companies wouldn't cover such a hazardous occupation as flying military aircraft, and the government did not insure nonmilitary personnel.[75] In several letters to her parents, Marie Mountain Clark, 44-W-1, poses questions about government insurance and makes inquiries about which private coverage might be available to her. Clark apparently assumed that the government insured the WASP. The question of her private insurance coverage was settled when her hometown newspaper ran a story about an emergency parachute jump she had made during a training flight. Clark's insurance company responded with a letter that enclosed a document for her to sign — a rider for "exclusion of protection for injuries sustained while on or falling from aircrafts other than those pertaining to passenger aircrafts on a regular passenger route." The Aetna Life Insurance Company offered congratulations in anticipation of Clark's completion of her WASP training course and — by the way — for her "return to earth via parachute!"[76] WAFS Evelyn Sharp's letters home also prove that she too never quite understood the benefits that pertained, or, more accurately, did not pertain, to her. Bartels notes that Sharp believed that she *was* insured by the United States military.[77]

Besides their lack of insurance coverage, the women were ineligible for other military benefits. In one training accident that took the lives of two people — a civilian flight instructor and a WASP trainee — the instructor's family received $10,000 from his insurance policy, and coverage of burial expenses, while Margaret Oldenburg's family received a $200 death benefit and a pine box, but no provision to transport her body home to California. Cochran covered those expenses herself.[78] In an interview, one former WASP states that a man killed in wartime service to his country could leave his family $10,000 from the government and the right to place a gold star in their window. However if the family of a WASP could not afford to transport her body, the women themselves collected money for the purpose. The woman's coffin could not be covered by an American flag, nor was her family allowed to display a gold star. The interviewee admits that those facts bothered her less during the time she served than they did later.[79]

Attending pursuit school in Palm Springs, California, WAFS Florene Miller waited on the ground for her turn with a flight instructor, who was working with her friend Dorothy Scott. Miller witnessed their fatal mid-air collision with a P-39. Miller notified Scott's family and she accompanied the hearse which carried her friend's body home to Burbank, California.[80] Fagan remembers another instance in which a dead trainee's classmates collected money to cover the cost of transporting her body home. Although a WASP representative accompanied each body on its final journey, officially there could be no flag, no casket or military funeral, and no insurance payment from the United States government.[81] WAFS Nancy Batson accompanied Evelyn Sharp's body from the site of her crash in Pennsylvania to her hometown of Nebraska for burial. Contrary to accepted practice for civilians, Batson gave the local funeral home permission to drape Sharp's casket with an American flag, an honor to which the dead pilot was not officially entitled.[82]

They may have been civilians, but the United States Air Force treated the women pilots as if they were military personnel. Elizabeth Strohfus was assigned to Las Vegas Army Air Field after graduation.[83] Strohfus says that she and her fellow WASPs took a military oath and were subject to the same regulations as male cadets.[84] Clark notes that WASPs were subject to military rules, such as the one forbidding them from fraternizing with "noncoms," noncommissioned officers.[85] However, although they worked within a military context and were expected to obey military orders, they were not actually subject to military discipline, a fact not lost on some of the women, who took full advantage of that situation.

Still, WASPs enjoyed some benefits derived from their quasi-military status. They had "number two" priority on domestic aircraft, meaning they could "bump" all passengers except the president of the United States; but for them, poor performance resulted in dismissal, rather than in reassignment or punishment as it did for male military pilots.[86] O. Vivian Fagan, in her memoir, *Zoot Suits and Parachutes: WASPs WWII*, tells a story about the time her WASP wartime "priority" bestowed on her and her companions a very helpful emergency tire repair in Gallup, New Mexico.[87]

Their uncertain status in relation to the army also affected their leaders. Both Love and Cochran were inexperienced in a military context. Adela Scharr was placed in charge of the WAFS contingent stationed in Romulus, Michigan. In her detailed history of the WAFS and the WASP, compiled from the hundreds of letters she wrote during the war, Scharr writes that Nancy Love decided that "changes [were] needed in the squadron" so she removed Scharr from her post at Romulus AAB and returned her to Wilmington, her original base.[88] Love's comments to Scharr about her subordinate's "unpopularity" and "selfishness," give one WAFS' perception of Love's management of her ferry pilots and it demonstrates how the leaders strategized as best they could within the unfamiliar military structure in which they functioned.[89]

Bypassing their early opportunity to become part of the United States Army — for purely practical reasons — meant that by the time a bill was introduced in Congress giving the WASPs military status, it was defeated. The women achieved veteran status only many years later when former WASPs made a concerted effort to fight for militarization in the late 1970s.

Compensation

Although performing identical work, some of the women pilots earned less than their male counterparts. Only in the Soviet Union, where a constitutional guarantee of gender equality existed on paper, if not always in practice, and during the later years of the war in Great Britain did women receive the same compensation as men for their service as military pilots.

When they first began, the women pilots of ATA earned only 80 percent as much as their male counterparts. The British treasury had ruled that women's compensation was to be 20 percent less than men doing the same work. ATA women were paid 230 pounds annually with 8 pounds per month flight pay.[90] After they were trained to fly all types of aircraft used by the RAF and there was no difference between the kinds of work they performed compared to their male counterparts, they were still earning less than ATA male pilots. Pauline Gower, who had advocated for advancing the Women's Section to higher classes of aircraft, discussed the matter of their unequal compensation with the minister of Aircraft Production, Sir Stafford Cripps. A few of the ATA's directors, with Gower's encouragement, also recommended that the women receive equal pay. Sir Cripps posed the question of women's compensation to the British treasury, asking for their advice on what he ought to reply to Gower concerning the women's lower pay rate. Because of his inquiry, the treasury ruling was changed in June 1943.[91] Veronica Volkersz notes that she too helped to campaign for equal pay for male and female pilots in the ATA organization. In her memoir, Volkersz writes that "After strong representation, including the intervention of Pauline Gower, the situation was adjusted."[92] This change in the British treasury ruling, giving the pilots of the Women's Section equal compensation, was a most unusual equivalency for that era and something that many of them specifically mention in their memoirs.

Volkersz cites the availability of insurance for ATA pilots as follows: 1500 to 2500 British pounds according to rank and type of airplane flown, figures she calls "not generous."[93] Helen Richey's biographer writes that the American women in ATA earned $4000 a year for their service, including their ATA pay, plus a $25 weekly deposit to their credit in the U.S.[94]

The 1936 Soviet Constitution, known as the "Stalin Constitution," promised universal voting rights to Soviet citizens as well as gender equality, including equal pay for equal work. In reality, the Communist Party determined whether these constitutional rights were granted. The Stalin era was a time of terror and Soviet citizens could expect purges and imprisonments rather than entitlements.[95] In Soviet society, women were accepted as a legitimate and necessary part of the workforce and it was common for them to perform heavy labor. Therefore no objections were raised, as in other countries, that the women who served in the Red Air Force would be "taking away jobs from qualified men."[96]

Soldiers in the Great Patriotic War earned relatively little, nothing more than "pocket money," according to *Daily Life Online*. While the Soviet population suffered persecution and famine, the Red Army endured privations as well. Although officers were paid more than regular recruits, they were expected to join the Communist Party as part of their career path and their salary was intended to cover the cost of uniforms, food, and furniture for their quarters. Living conditions were so bad that morale in the Red Army was a chronic problem for the government. Stalin's purges, from which the military was not exempt, created a pervasive climate of suspicion and paranoia.[97]

In the United States, WAFS earned $3000 per year and received $6 per diem on ferrying trips.[98] WASP trainees were paid $75 twice a month. Although their flying clothes, textbooks and bed linens were furnished, they bought their own regulation gym clothes.[99] The women's base pay was higher than that of their male officer counterparts, but their status as civilians deprived them of any government benefits pertaining to military personnel. The Wings Across America website, using statistics gathered by WASP Byrd Howell Granger, whose detailed history of the formation and service of the WAFS and WASP helped prove that the WASP had indeed performed military service, cites the women's base pay, when stationed at an air base, as $250. They did not receive housing, nor did they receive flight pay or a uniform allowance.[100]

Numbers

In Great Britain, there were 166 members of the Women's Section of the ATA, a little more than one tenth of the total number in the organization, writes Lettice Curtis. Thus women represented a fairly significant percentage of the entire group.[101]

The initial flight group formed by Marina Raskova yielded three regiments whose numbers amount to somewhat less than one thousand women. In the Soviet Union, most regiments had two squadrons, each having ten aircraft. The regiments included pilots, navigators, mechanics, armorers, and other ground crew in different proportions depending on the nature of the work to be done. Pennington, in *Wings, Women and War: Soviet Airwomen in World War II Combat*, gives the following breakdown of the 745 women in the three aviation regiments: 46th Guards Regiment: 261; 125th Guards Regiment: 300; 586th Regiment: 184.[102]

Of the women that Nancy Love initially contacted, 13 were accepted into the group of WAFS "originals." By 1943 the number had grown to 27, and ultimately the number of WAFS settled at twenty-eight.[103] Only 134 women flew fighter aircraft for the American Air Transport Command, a number less than one-fourth of those who flew all other types of military aircraft for the United States during the period.[104] A few more than one thousand women pilots flew for the United States Air Force as WASPs. Betty Stagg Turner's account, *Out of the Blue and into History*, lists eighteen WASP classes which graduated in the years 1943 and 1944, and she includes biographies of twenty WAFS participants. Of the 25,000 applicants to Cochran's training program, 1830 women were selected. From that group, 1074 women earned their wings and actually served as WASPs.[105]

Publicity

The formation of a Women's Section in Britain's ATA was a news sensation and the organization's first eight women were subject to extensive press coverage. They were the recipients of both adulation and criticism. Some members of the public claimed that the women would take valuable employment away from men. That perception was especially strong during the "phoney war" in Britain, a time after England's declaration of war in September 1939, but before the country experienced any wartime action. During this time there was intense competition for war jobs because the British economy was still suffering from the Great Depression. Gower informed the public about her pilots through articles, speeches, and radio interviews. Her objective was to allay fears that the women would siphon off valuable aviation jobs and deprive family men of a good employment opportunity. At the same time, as their leader, Gower stressed to the women the critical importance of performing their work with care and absolute efficiency to counter negative public opinions about the abilities of females to fly military airplanes. Here are Gower's words from one radio broadcast: "We are a small group of women pilots with a job to do. We are just helping, along with others, to win the war. Our job will not be obtrusive. But it is going to be well and efficiently done."[106]

Gower was not averse to allowing the first members of the Women's Section to be interviewed and photographed, but she wanted to control the amount and character of the publicity they received. Newspaper accounts weren't always accurate, especially when the press staged the scene. For example, the photo of the first eight women in their new ATA uniforms enjoying tea while seated at linen-draped tables beside an airfield was unrealistic, to say the least. The headline "Eight girls will 'show' the R.A.F." is also typical.[107] Gower also coped with the pressure that the women felt to perform perfectly, lest they prove their critics right, according to Rosemary Rees du Cros.[108] Curiously, the South African "Jackie" Sorour Moggridge, who joined ATA as its youngest member in 1940, states that it was imperative for the existence of the Women's Section to be kept a secret in its early days because the British military didn't want Germany to think England was so desperate that it had resorted to using *women* as pilots![109] In light of Pauline Gower's openness to the press when she assembled her group, Moggridge's statement is puzzling, but it shows how the women's perceptions could vary, even within the same organization.

In the Soviet Union, news that women's regiments were forming was disseminated unofficially by word of mouth and officially by the Komsomol, which contacted Soviet flying clubs and technical schools in its search for volunteers. The existence of women's aviation regiments was not widely publicized in the Soviet Union either during the war or later. They were not featured in the domestic or the foreign press, so few print or broadcast items exist covering the women's aviation regiments. When they were included in articles, the women's names were printed using their masculine rather than feminine forms. In the Russian language, there are masculine and feminine forms of surnames, for example, Marina Raskova's surname had derived from the fact that she had married a man named Sergey Raskov. In time, the women's regiments — particularly the 46th — were covered in the "front-line newspapers" published by the Red Army, although the 125th Guards Bomber Aviation Regiment continued to be overlooked.[110] Although Joseph Stalin knew Raskova personally and had given her permission to form and train the women's aviation group, he never publicly mentioned the women's contributions either during or after the war. In his speeches he acknowledged the urgency of rapid

Soviet preparations for war after Germany's attack, and praised the continuing valiant effort of the entire Soviet population.[111] In speeches celebrating the end of the Great Patriotic War, he credits female partisans and collective farm workers for their wartime work, but he makes no mention of women's service in aviation regiments.

In the United States, the general public was unaware of the existence of the training program for women pilots, except for a few people living near the air bases where the women were housed. Indeed, the existence of units for women in the U.S. Air Force wasn't even well known in the American military itself. A *New York Times* article had announced the formation of Love's new squadron of female ferry pilots, but the creation of Cochran's WFTD was kept quiet, at least initially, at her request. Cochran's desire to keep her experimental training program unpublicized spawned rumors among people living nearby who wondered why there were so many young women suddenly appearing in Sweetwater.

One of Cochran's requirements for her trainees, which resulted in confusion among the public and even within the military itself, was that they must keep the existence of their flight school a secret. Reportedly, Cochran instructed the new arrivals to say anything they liked in response to people's inquiries, but *not* that they were learning to fly the army's airplanes.[112] Her wish for discretion lasted well into the life of the program, as Doris Brinker Tanner, 44-W-4, notes in *Zoot-Suits and Parachutes and Wings of Silver Too!: The World War II Air Force Training of Women Pilots, 1942–1944*. Tanner writes that WASP executive officer Deaton welcomed her own class with the following words: "Do not inform your families of any facts or figures concerning what you are doing, inasmuch as such information is classified and you are under regulations forbidding you to do so."[113] Helen Harrison, the Canadian who flew in the ATA, recalls that the formation of women's air force units in America was not well known among Canadian women pilots either.[114]

When the first class was about to graduate, Cochran had little choice regarding publicity, as news organizations converged to cover the highly unusual event, which gave them the opportunity to snap a photo of Admiral L.C. Colbert pinning wings on his daughter, Mary Lou, during the ceremony.[115] Articles in newspapers and magazines featuring the WASP were typically so contrived as to fall well short of accuracy either about the women or the work they were doing. One wartime newsreel juxtaposes film of a WASP trumpeting reveille against a scene of sleepy trainees turning off their alarm clocks and rolling out of army cots.[116]

4

Becoming Military Pilots

All the women accepted for service followed military orders, even when they were civilians. Their presence on military air bases was largely unprecedented and their roles in military aviation were new to them, to their leaders, and to nearly everyone else that they met in the course of their daily work. The fact that the Soviet women served in regular combat units in the Red Air Force gave them a status, though highly temporary, that was also quite rare. They all learned about the aircraft types used in their countries and they absorbed the basic routines of army life, including obeying rules and regulations, and marching in formation. The duration of their training period varied, the major determinant being the urgency of their country's need for pilots. Whether they served as military personnel or not, except for the American women's initial days of service, all the women pilots were issued uniforms and equipment appropriate to the work they would perform and to the variety of flight conditions they would encounter. The fact that they occupied a formerly all-male context meant that most of the work clothing issued to them was several sizes too large because it had been developed to fit an average male. The only women who escaped this inconvenience were those in the British ATA Women's Section, for whom uniforms were immediately available. The women's uniforms sported distinctive patches and emblems, as well as those of their country's air force. Some of the Soviet women earned special honors which gave them the right to wear medals and uniforms pertaining to their valor in battle.

Although the women were in a distinct minority wherever they served, life was not always smooth, even within their own aviation units. Rivalries developed, often stemming from the clashes of strong personalities. Sometimes unavoidable circumstances exacerbated these tensions. Few of the women who flew military aircraft during the Second World War had had the training necessary for performing effectively as military pilots. Consequently, the new recruits' first assignment was to report for duty at the air bases where their ground and flight training would take place.

Training

Great Britain

In the relatively compact island nation of Great Britain, long travel distances were less common than in either the Soviet Union or the United States. The first ATA women traveled from their homes at their own expense to Ferry Pool No. 5, Hatfield Aerodrome, northeast of London, to take their flight tests.[1] Veronica Volkhersz writes that when she received her ATA letter inviting her to report for a flight test, she "hired a car" and drove to Hatfield from London, where she had been working as an ambulance driver.[2]

For a few other members of the Women's Section, the journey was much longer. A number of them made a dangerous ocean crossing in order to fly airplanes in England. These women included the small contingent led by Jacqueline Cochran, a few Canadians, including Helen Harrison, the Chilean Margot Duhalde, and Jackie Sorour Moggridge, who travelled from South Africa just before the start of hostilities greatly increased the dangers of ocean travel. Anna Leska and Barbara Wojtulanis escaped, along with members of the Polish air force, before their country was invaded by Germany. Leska, who had been a pilot officer in Poland, flew to Rumania, where she and her companions were captured. With the help of a Polish squadron leader, she escaped to Hungary and then trekked through France, eventually finding a boat that could take her across the English Channel to Plymouth.[3]

Although they were already competent aviators, the first eight members of the Women's Section of ATA needed instruction in military aviation. The BOAC flight school provided courses for all ATA pilots at the White Waltham Aerodrome near London. Diana Barnato Walker, who had relatively little flying experience before she was accepted into the ATA somewhat later, credits the excellence of its pilot training for the organization's "amazing" efficiency.[4] The pilots were instructed in the subjects of meteorology, balloons surrounding towns and factories, map reading, signals, technical data, engines and navigation, according to Walker.[5] In ground school they learned about British aircraft models and their engines, how to use Morse code and the Verey (flare) pistol, and how to recognize and interpret various signals. After the trainees had finished their ground classwork, they were ready for dual flight instruction, and then they soloed.

The ATA emphasized cross-country training; but without radios, navigation was a matter of making careful calculations using compass and map. The pilots were required to make a number of cross-country flights, which helped orient them to the often unfamiliar English terrain. Before they began ferrying airplanes, even experienced pilots were required to complete these navigational flights in relatively slow airplanes known as Moths and Magisters. Roberta Sandoz Leveaux, one of the Americans, found this particular aspect of orientation extremely valuable. It not only helped her to learn about the English landscape, but, because she was accustomed to the sunny American West Coast, it also gave her much needed experience in coping with changeable weather conditions. According to Leveaux, the ATA instruction program required twenty-five such flights. Leveaux recalls that the ferry pilots were "paid to be safe, not brave."[6]

ATA training stressed safety above all else, because the airplanes had to be delivered in good shape regardless of conditions. After successfully completing their training, the pilots received their wings. As the war went on, ATA training was provided for women with no previous flying experience at all. The washout rate for the ATA Women's Section — the British term is "wastage rate"— was just under 5 percent higher than the rate for ATA men, according to Canadian Vi Warren Milstead, as quoted in Render's *No Place for a Lady: The Story of Canadian Women Pilots, 1928–1992*.[7]

Soviet Union

On October 10, 1941, a few months after the German air attack, Soviet troops had pulled back to a point only fifty miles outside the city of Moscow.[8] Moscovites expected an invasion by advancing German troops at any moment. Many of the women whom Raskova chose for her flight group had recently been digging trenches outside their city in an attempt to forestall a ground attack. Her flight group assembled in Moscow, but it became necessary to relocate the women to where their training could proceed without the threat of attack. The outlook for Moscow was dire; trains headed east away from the city daily and one of those trains, consisting of freight cars that also moved soldiers, transported hundreds of women in the flight group to a large training base at Engels on the Volga River, about five-hundred miles to the southeast, a less immediately imperiled place for the women to train. Their journey lasted nine days, its duration not entirely attributable to the distance but to the fact that the train needed to stop periodically to let more critically needed shipments through on the same track. When they left Moscow, the temperature registered below zero Fahrenheit. Necessities like food, water, and heat were in extremely short supply. Raskova traveled on the train with her volunteers and spent time in each of their freight cars, thus demonstrating her interest in the women's welfare, an action that some of them later said served to inspire them for the task ahead.[9]

While they trained at Engels, the Soviet women lived in a former gymnasium, in barracks similar to those of other military personnel, and they endured cold weather conditions that were only intensified by the open plains surrounding the airfield. As befitted her status as a commander, Raskova was offered a more comfortable room than those her volunteers occupied, but she eschewed any special accommodations offered on the basis of her rank and insisted that she and the other commanders share a room located in a former barbershop, a place that was sometimes so cold that the women sought shelter with their new flight group instead.[10] Nevertheless, the new trainees often sang popular songs or recited poetry. Although winter was well underway by the time they arrived, they flew open-cockpit biplanes, Po-2s constructed of plywood, in which they practiced loops and other air tactics and fired at sleeve and ground targets. The Po-2 would be flown throughout the war by the 46th Guards Aviation Regiment, the pilots known also as "night witches."[11]

The women learned theory, armament, and other military subjects previously unfamiliar to them, although many were experienced and competent pilots and some had been instructors in civilian aviation programs. They learned how to fly in formation, a necessary tactic for defense against the enemy in battle conditions. This involved having several airplanes take off simultaneously in formation, so that the additional airplanes could serve as wingmen while all carried out the orders of the formation's leader.[12]

Soviet women received training equivalent to that provided for men because there would be no distinction in the kinds of work they would perform compared to their male equivalents, including flying in combat. The usual flight training program was accelerated at Engels, condensing three years of flight instruction into six months. Raskova designed and oversaw the entire training program for all the Soviet women's regiments, a responsibility that kept her directly involved in instruction programs operating twenty-four hours a day.[13] In an unusual concession, The Red Air Force had allowed Raskova to form a group of female pilots on the condition that she alone would be in charge of their training. This included also providing instruction for ground support staff, flight mechanics and armorers, all of whom would be women.[14]

At Engels the women became accustomed to military discipline. The first order they received upon arrival was to get a "boy-style" haircut, leaving them all sporting a few inches of hair at the forehead with closely shaved sides and back.[15] Olga Yakovleva, a member of the 586th regiment, recalls that among the first things that happened to the women was losing their plaits (long braids) and getting identical "tomboy" hairdos. For most of the women, having short hair was only a brief condition. They let their hair grow after training was over. Yevghenia Andreyevna Zhigulenko, from the Night Bomber regiment, explains: "First we had cut our hair like boys, but then we let it grow long. It wasn't authorized, but what were you going to do with pilots? They looked death in the eye every day, so nobody could touch them."[16]

Yakovleva remembers that Raskova was a demanding commander and did not indulge the women at all.[17] Pennington quotes one volunteer who describes Raskova's punishment of a few trainees who, in an attempt to gain a few extra minutes of sleep one night, tried to cheat by wearing nightshirts under their overcoats when summoned to appear at attention. When Raskova discovered their deception she ordered them to march around the airfield bare-legged in the cold wind.[18] Their commander knew that in battle conditions, taking shortcuts and ignoring orders would put their lives at risk.

Statistics on retention rates for the trainees do not seem to exist. New volunteers for the aviation regiments continued to arrive at Engels while the original flight group trained. The newcomers petitioned Raskova for acceptance, but not all the women who showed up, sometimes on foot and often without money or food, were retained.[19]

United States

In the United States, the women recruited by Love or Cochran paid their own travel expenses to their various training bases, each one choosing her most convenient or affordable mode of transportation. Women accepted into the WAFS journeyed to Wilmington, Delaware, for training. Those accepted into the WFTD made their way to Houston and later to Avenger Field in Sweetwater, Texas. Tanner describes her arrival in Sweetwater in November 1943 as a new trainee in class 44-W-4, the culmination of a long train ride from Brownsville, Tennessee: "The train screeched to a grinding halt with a jolt. I pushed my carefully packed, new, brown Samsonite bag out into the aisle, pinned the felt pill-box on back of my head, pulled on my white cotton shorty gloves, stuck my purse under my arm, grabbed my cosmetics train case and made my way down the aisle and steps."[20]

The highly experienced pilots chosen by Nancy Love for the WAFS received four weeks of army training, somewhat longer than the length of training provided for their male equivalents.[21] The WAFS' sole responsibility would be ferrying airplanes cross-country. Their training consisted of an orientation to the military, and learning how to fly "the Army way." The story of Teresa James is typical. James, who began flying in 1933, was a flight instructor in America's Civil Air Patrol when she received a telegram inviting her to join Love's group of women pilots and begin her training in Wilmington,

Delaware. James found that her new WAFS flight instruction helped her to lose the bad habits she had acquired as a natural response to putting an airplane down on small, rough, frequently tree-lined airfields, where she had first learned to fly. In *On Wings to War: Teresa James, Aviator*, James recalls that the army demanded absolute accuracy of technique.[22]

After Love's WAFS and Cochran's WFTD merged to become the organization known as WASP, the women's flight training program could be designed according to Cochran's specifications. The training program now lasted six months and it generated pilots who were able to perform other tasks besides ferrying. WASP class sizes averaged about twenty-five, with additional women reporting for training each month. At Sweetwater, each class of trainees was divided into two flights, groups that had complementary daily schedules designed to make optimum use of Avenger Field's available resources. Half the class attended ground school in the morning while the other half practiced flying. In the afternoon, the assignments were reversed. Each week the morning flight assignment changed.[23] The women received the same training as male cadets, except for gunnery, which was omitted from their curriculum because they were never expected to use that skill. They would be excluded from combat missions.[24] The flight training course included close-order drill and the physical training necessary to improve upper body strength which the women required to successfully handle large military aircraft, the types Cochran wanted them to fly.[25] Trainees also learned to fly at night. Tanner, in *Zoot-Suits and Parachutes and Wings of Silver Too!*, notes that their ground school consisted of twenty hours of mathematics, twenty hours in theory of flight and aircraft, forty-two hours in engine and propellers, twenty-four hours in physics, fifty hours of meteorology, eighteen hours of instruments, seventy-six hours of navigation, and thirty hours of Morse code, in which they were tested periodically so they would maintain proficiency at the level of ten words per minute.[26] The WASP training program generated a much larger group of pilots than the WAFS had, and they were able to perform many different kinds of flying tasks.

The story of one of the WASP flight instructors shows the circuitous path some of the women trod in the program. Dorothy "Dot" Swain Lewis, 44-W-5, had been in a pilot training program in Tennessee started by aviator Phoebe Omlie.[27] Omlie had been an important voice in the Bureau of Commerce Air Marking Program, pushing for the involvement of each state in the air marking program and directing the all-women staff.[28] Prewar flight training programs created a need for more instructors, so the Tennessee Bureau of Aeronautics at Gillespie Airfield in Nashville started the Women's Research Flight Instructor School in September 1942.[29] Lewis was one of ten women in that experimental program, which was designed to determine whether women could be trained as civilian flight instructors. When the ten graduated, the experiment was deemed a success and the training of women instructors was presumed to be the government's responsibility from that point on. In *How High She Flies*, Omlie's trainees, all in identical light pants with dark belted jackets, white blouses, and matching shoes, are pictured watching as she demonstrates how to clean up the kitchen.[30]

After graduating from Omlie's flight school, Lewis instructed U.S. Navy air cadets in Portales, New Mexico, but hearing her friends' stories about the WASP convinced her to join. Lewis became a flight instructor for class 43-W-8, one of the few female instructors in the program. The others were Helen Duffy and her sister, Dorothy, and Jennie Gower. Lewis tired of instructing and eventually resigned so she could enter class 44-W-7, but her level of expertise allowed her to advance to 44-W-5.[31] WASP trainees occupied barracks at Avenger Field, and the rules forbidding male instructors and female trainees from fraternizing made it possible for Lewis and Gower to maintain separate accommodations in a nearby house. Lewis considered the rules against trainees visiting instructors' housing to be "quite silly" in her case, and she and her housemate hosted the trainees at several get-togethers.[32]

O. Vivian Fagan, in her book, *Zoot Suits and Parachutes*, remarks that Sweetwater's Avenger Field was the only all-women training base in the United States, and consequently, it was the only American airfield that offered the full range of flight training — primary, basic, and advanced — in the same location, a situation that Fagan later realized was an unsafe mix. Fagan claims that the highest number of "wash-outs" occurred during basic training.[33] However, the elimination rate for women was lower than that of male trainees. Of the 1830 women who were accepted into the WASP training program, 58.7 percent graduated, according to statistics gathered by Cochran for her *Final Report on Women Pilot Program*.[34]

Because militarization for the WASP was considered imminent almost to the end of its existence, Cochran arranged for several women to attend officer training, a practice consistent with the career path available for male pilots. Marie Mountain Clark was among the students in officers training school (OTS) in Orlando, Florida. In *Dear Mother and Daddy: World War II Letters Home from a WASP, An Autobiography*, she lists her courses as follows: army orientation, air force administration, military courtesy and discipline, base and staff functions, group, post, and squadron duties, AAF communications, military law, AAF weather, recognition of aircraft, organization of war departments and AAF, air force supply, safeguarding military information, chemical warfare, and aero-medicine.[35]

Jacqueline Cochran speaks at the graduation ceremony for class 43-W-1 at Ellington Army Air Field in Houston, Texas. WASP uniforms were not yet available. Graduates wore a semiofficial uniform of white shirts, tan pants, and overseas caps (The Woman's Collection, Texas Woman's University).

Knowing that her experimental program would always be held to more scrutiny and higher standards than an equivalent group of males, Cochran chose Mrs. Leni Leoti "Dedie" Clark Deaton to assume the role of executive officer, which meant that Deaton lived among the trainees as their "housemother." Her first challenge was finding accommodations for them in Houston because the training base was not prepared for the arrival of females. Finding few alternatives, Deaton asked Cochran to approve her request to house the women in a motel called Oleander Court, but the director of Women Pilots was extremely reluctant to place young women from her brand new program into a place like a tourist court because such facilities had yet to outlive their unsavory reputations. However, the compelling need to find the women accommodations *somewhere* combined with a lack of available options and gave Deaton justification for overriding Cochran's objections.[36]

Cochran's early life experiences tended to affect her management of the women under her command, as the above example shows. Hints of the self-protective strategies she had developed as a young girl while working alongside grown men in itinerant Florida logging camps are constantly evident in the importance she placed on respectability and her sensitivity to behavior she considered inappropriate. Cochran's insistence on high standards in the women's performance *and* their behavior was absolute, not only because her program's survival depended on the public's perception, but also because the WASP program was so clearly linked to her own name and reputation. When the American women arrived in England, she forbade the ATA's physician to require *her* pilots to fully undress for their physicals, a practice accepted as standard procedure across the Atlantic. Most people were unaware of the circumstances Cochran had faced as a child, so there were many who considered her objections "prudish."[37]

One anecdote can illustrate the sort of work Deaton performed as "housemother." WASP trainee Elizabeth Strohfus loved to fly the AT-6 advanced training airplane. When her hometown boyfriend visited her in Sweetwater, he begged her to marry him and forget "this nonsense." Strohfus seriously considered returning home with him; but she delayed her final decision, claiming that she needed time to say good-bye to her friends—and to the airplanes. When Deaton learned that Strohfus was about to resign because her boyfriend wanted to get married, she asked her, "What about you?" Deaton cited her excellent record and ordered her, before resigning, to take one last flight in an AT-6. Afterwards, Strohfus phoned her boyfriend to say "I'm not coming home!" He replied, "[I]f you don't come home, I'm going to marry someone else." She answered, "[Y]ou go right ahead." Strohfus writes that they both lived "happily ever after" when he did.[38]

Tanner, in *Zoot Suits and Parachutes and Wings of Silver Too!*, draws on Deaton's private files to describe events missing from other accounts, including the text of an undated letter from Deaton to Cochran concerning a WASP trainee and a recent graduate who appeared to have been involved in a lesbian affair at Avenger Field. Deaton duly reported the facts, which were supported by testimony from the women's classmates and her own investigation. Deaton wrote that she was "sickened" by the fact that "we had let a girl graduate and go out from here to represent us who is unquestionably abnormal morally." This incident ended with the dismissal of two trainees in Sweetwater. Deaton's investigation also found that the trainee left behind when her lover graduated had then begun a relationship with another. Deaton writes that "[she] impressed on the girls the necessity of reporting any irregularity before it goes beyond that. They liked one of the girls very much and I poured it on them pretty hot about not reporting the first two before the other got involved — reminded them that they as part of this program would be judged by the girl who has graduated and is out representing them, and also that the other girl might never have got involved if they had told us before the first girl graduated."[39] Deaton also informs Cochran of a trainee who was sent home after attempting suicide and another who was dismissed for getting drunk. In response, Cochran visited Sweetwater early in 1944 and she also stopped at the (unnamed) army air force base in California where the above mentioned graduate was stationed.

Housemother Deaton was also an advocate for the trainees. She instituted a review board consisting of the commanding officer, the flight instructor, the check pilot, and either herself or her assistant for each trainee who faced "washing out" after a check ride.[40] She had observed instances in which a check pilot whose advances had been rejected by a trainee would fail the woman in retaliation. The review board served to double-check against those responses.

Although the first WFTD trainees in Houston had temporary accommodations in local hotels, once they moved to Avenger Field, the women were housed in barracks. During their training period they had ample time for relaxation. While training in Sweetwater, Clark wrote to her parents requesting money or personal items, such as phonograph records or magazines like *Mademoiselle*, because she and her classmates enjoyed dressing up on weekends.[41] She adds that the young trainees weren't always serious and sometimes they knowingly broke the rules. Cochran admonished the women at one WASP graduation ceremony, warning them not to "hedge-hop," that is, fly while barely clearing trees or other obstacles and pulling up at the last minute, a warning that came too late for most of them, writes Clark.[42]

Trainees at Sweetwater devised their own words to well-known popular tunes. They learned the Army Air Corps Song and made up the song "Zoot-Suits and Parachutes and Wings of Silver Too," sung to the tune of the popular song "Bell Bottomed Trousers." Doris Tanner found that singing on their way to ground school, the mess hall or while riding in the bus which transported the trainees to the airfield, raised their spirits during the rigors of training.[43] Two newsletters, the *Fifinella Gazette* and the *Avenger*, each comprising five issues, were published by WASP trainees in 1943.[44]

O. Vivian Fagan describes the "Eager Beaver" show she and her friends created during their training and which Mrs. Deaton encouraged them to take "on the road"—the tradeoff was that they had to abandon their flight training. When they realized that, the amateur performers declined.[45] Fagan says that trainees spent their spare time at nearby Sweetwater Lake. Trying to deepen their tans, some of them removed their shirts while flying, an opportunity that male pilots in the area couldn't ignore and they tried to fly close enough for a glimpse. Swarms of locusts were part of life in Sweetwater and the women also experienced the frequent sandstorms of windy west Texas. Although officially forbidden to do so, after the nightly bed check trainees moved their cots outside their barracks on hot summer nights.[46]

Uniforms, Equipment, Emblems

Great Britain

When she founded the Women's Section, Pauline Gower arranged for her pilots to have their own uniforms, which were identical to those worn by male ATA pilots except for their matching skirts. Skirts were highly impractical for activities like flying; but consistent with the custom of the era, the women rarely wore pants. Lettice Curtis, in *The Forgotten Pilots: A Story of the Air Transport Auxiliary, 1939–45*, remembers flying in a skirt. In fact, Curtis reports that most of the women she knew didn't even *own* a pair of pants. Official ATA Women's Section uniforms included a navy-blue belted tunic with four pockets, one skirt and one pair of pants—to be worn only on the air bases. There was also a great coat and forage cap. The women were required to purchase black stockings and shoes, a black tie, and blue RAF shirts. Milstead notes that the women's uniform was exactly the same as the men's except for the skirt, which was worn with black stockings for ceremonial occasions. The women's hair was cut above the collar and their shoes had to be low-heeled.[47] Tunics add one gold stripe for a second officer and two gold stripes for a first officer. Pilots' uniforms had gold embroidered wings. Milstead describes the several ranks in the ATA, beginning with cadet (training) and ending with commodore, the title held by ATA's commander, Gerard d'Erlanger. Curtis notes that the first eight women also received sheepskin jackets and pants for warmth in the air while they flew in open-cockpit aircraft. Although she says the sheepskin outfits were too bulky to be considered practical, the women often used their leather flying boots in winter.[48] Harrison adds that when ferrying single-seater airplanes, the aircraft's hood needed to be open during takeoff and landing, so besides their regular uniform jacket and slacks, the pilots wore coveralls, goggles and a helmet.[49]

The ATA women were fitted for uniforms at a tailor in London so as to be suitably attired for their publicity photographs. However, the tailors, who were used to servicing men only, were reluctant to take a woman's measurements, their hesitation yielding predictable results: the women's ill-fitting trousers almost always needed alteration.

Soviet Union

The Soviet women were issued the same clothing as male soldiers in the Red Army. This clothing packet even extended to men's boots and underwear. The women stuffed extra material into their new footwear to try to minimize the over-sized fit.[50] Not until 1943 did the women in Soviet aviation regiments receive their own uniforms, instead of the ill-fitting male uniforms originally issued.[51] Lidya Litvyak, during training at Engels, removed the fur lining from the boots issued with her winter uniform and sewed it onto her uniform's collar. During the following day's inspection, Marina Raskova noticed the unauthorized addition and ordered the young pilot to spend the next evening removing her handiwork.[52] When two of the three women's regiments had earned the honorary title of Guards, they were issued special Guards uniforms along with accompanying medals.[53]

United States

Nancy Love selected the design and the gray green gabardine fabric for a new uniform, consisting of a belted jacket with a collar and pockets, which was worn with a matching skirt or pants. The WAFS uniform displayed the Air Transport Command insignia on the left breast pocket, to which were added gold wings and the U.S. Air Force insignia. An overseas cap completed the outfit. The women also received cold-weather flying suits.[54] Because they were civilians, the WAFS paid for their own uniforms.

The problematic British uniform tailoring experiences recurred for the American women when they were fitted for their uniforms. WAFS Teresa James describes the measuring process in Wilmington for her first uniform pants at Carlson's Tailor Shop: "We were standing up and Mr. Carlson was on his knees. He measured the outside and instead of coming around with the tape to measure the inseam, he just looked. He guessed the inseam measurement, as he was too embarrassed to touch a lady to measure the inseam's length."[55]

4—Becoming Military Pilots 65

At first, WFTD and WASP trainees were issued men's flight suits, which the women called "zoot suits" because they were sized for the average 42–44 male. The suits hung in voluminous folds from an average-sized female frame, resembling the well-known popular men's attire of the 1930s. The women adjusted their new clothing by pulling the fabric up under their arms until the crotch approximated its correct location and tightening the belt so the suit folded over at the waist. Then they rolled up the sleeves and pants, according to Tanner.[56] On training flights the women were ordered to wear light-colored head wraps called "Urban's Turbans," named for the WASP commanding officer, Major Robert K. Urban, who thought the headgear would keep the women's long hair from flying in the faces of their instructors in windy open cockpits.[57] Tanner remembers that upon arrival in Sweetwater, the director of flying told trainees that they must get their long hair out of sight: "[W]ear your turbans or hair nets or better still, cut it off."[58] In the film *A Brief Flight*, one former WASP says that working clothing was never issued to them, only a dress uniform. Another says that the WFTD were also issued men's coveralls and leather flight jackets, but they had to purchase the rest.[59] Some of the women recall washing their "zoot suits" in the shower. In summer they dried quickly in the Texas heat.

After the groups' merger, Jacqueline Cochran determined the style and color of the new WASP uniforms, including where they would be designed and manufactured. The typical uniform included a skirt, trousers, jacket, coat, hat and gloves.[60] The color was different from the gray hue chosen by Love because General Arnold anticipated that the air force would soon become a separate and distinct organization from the U.S. Army where it had originated as the Army Signal Corps. Arnold told Cochran that he wanted the uniforms to be blue.[61] Betty Greene, 43-W-5, agreed that the WASP uniform was designed to match those of the soon-to-be-established air force.[62] When a few prototype uniform designs were ready, Cochran presented them to General Arnold, but she made sure that the design she herself favored was worn by an attractive (professional) model. Photos depicting women wearing WAFS uniforms alongside women attired in WASP uniforms indicate that Love's "grays" weren't abruptly abandoned in favor of Cochran's "Santiago blues." Because their official WASP uniforms weren't yet ready, the first graduating class in Houston wore khaki pants and white shirts for their ceremony, the "semiofficial" uniform. Cochran wore a flowered summer dress. The first class that wore WASP uniforms was 44-W-1 at their graduation on February 11, 1944.[63]

The WASP mascot, Fifinella, appeared on the gateway to Avenger Field as well as on the uniforms' patches. She was a character whose origins can be traced to early legends of gremlins among British RAF pilots who were stationed in India in the 1920s.[64] In *Women with Wings,* Pauline Gower writes about flying to Scotland, a place she calls "gremlin country," rugged and misty terrain inhabited by scissor-wielding elves intent on cutting the wires that held biplane wings together.[65] Hapless pilots could lose control attempting to evade gremlins and crash into unseen hillsides. Former RAF pilot and children's author Roald Dahl was injured during the Second World War and then reassigned to the United States as a member of Britain's wartime diplomatic corps in Washington, D.C. Dahl's unpublished manuscript about gremlins caught the attention of Walt Disney, who purchased rights to the story and published it in book form in April 1943.[66] Disney's *The Gremlins* contains the first reference to "fifinellas," wingless female gremlins who became the inspiration for Disney's design for the WASP mascot.

Hostility Without—

Great Britain

Having ferried aircraft long enough to develop an excellent safety record, the women of the ATA were at times still "unpopular," as Gower phrased it.[67] No matter how skillful they became, there were those who remained skeptical of the abilities of women to manage military aircraft. The initial resistance to a Women's Section was strong among RAF pilots and male members of the ATA alike. However, the need to prove their worth in a traditionally male occupation doesn't seem to dominate the consciousness of most of the women who served, judging from their memoirs. Indeed, Curtis claims that little prejudice existed within the Air Transport Auxiliary, and she speculates that the high level of acceptance for women pilots might have been the product of the tolerance of key administrators; or it may have been an unexpected benefit of the "mature" nature of ATA pilots, seasoned aviators who for one reason or another were not eligible for the RAF.[68]

Rosemary Rees du Cros writes in *ATA Girl: Memoirs of a Wartime Ferry Pilot* that the first all-women ferry pool established at Hamble was sometimes called the "lesbians' pool," though she didn't think anybody there actually was. One of the two male pilots still stationed at Hamble in the early months of the women's tenure even conducted a test of his own, bringing a mouse into the ferry pool one day, saying he just wanted to see "if any of you really *are* women." Du Cros reports that only one woman met his expectations by jumping onto a chair!⁶⁹ Du Cros goes on to say that both sexes in the ATA did the same work, held the same ranks, received the same pay, wore the same uniforms, and endured the same conditions, adding that "chivalry was neither expected nor offered."⁷⁰ Near the end of her memoir, du Cros muses on the fact that she detected little prejudice against her as a woman, even while performing her unusual wartime work. She writes that "The pretty little barriers that are put up in peacetime to make society more pleasant melt away in a big crisis and life becomes a grim struggle of tired grey people all doing whatever it is they can do to survive."⁷¹ Not every woman concurred. Jackie Moggridge notes several instances of prejudice against the women during her interview for the film *WASPs and Witches*. Still, on one occasion, she was invited to fly along on a combat mission with RAF pilots, a memorable experience which Moggridge recalls as more exciting than terrifying because it did not "feel" real.⁷²

References to its female participants appear throughout Cheesman's *Brief Glory: The Story of the ATA*, a postwar history which describes Lettice Curtis' first solo flight in a Halifax. She executed a perfect landing, at which the station master exclaimed, "And my lads have always kidded me how difficult Halifaxes are. Why damn it, they must be easy if a little girl can fly them like that!"⁷³ When informed that Curtis was actually *not* so little and that she had 2,000 hours and many types of aircraft listed in her logbook, he expressed surprise before departing for his pre-lunch drink.

Veronica Volkersz did not enjoy her assignment on the ground with the Women's Auxiliary Air Force before she joined the ATA because she felt that the "Waafery" gave a chilly reception to women pilots because they were jealous.⁷⁴ In *The Sky and I*, Volkersz describes the reception she and others received from the men stationed at Whitchurch, formerly an all-male ferry pool: "What? Bloody women being posted here? If they think we are going to run around carrying their parachutes, they have got another thing coming."⁷⁵ Volkersz claims that eventually she and her cohorts "got [the men] quite tame," to the extent that late in the war, her commanding officer assigned her to deliver a jet aircraft, the Meteor III EE386, an unexpected privilege for any pilot.⁷⁶ Overtly prejudiced males made fertile ground for jokes. Volkersz includes the following anecdote that circulated widely. Jackie Moggridge, the youngest ATA member, had delayed her takeoff in the first Albermarle she'd ever flown one day to allow a distinguished air marshall needing immediate transport to London to dash on board. Only after they were airborne did he notice that his pilot looked like a schoolgirl. Moggridge unsettled him even more when she waved her pilot's notes at him and confided, "You know, I've never flown one of these things before!"⁷⁷

Soviet Union

Women were not unknown in the Red Army prior to The Great Patriotic War, so when Marina Raskova formed her women's aviation group after Germany invaded, its volunteers were treated the same as other military personnel and they were expected to perform identical work. In time, male pilots were assigned to two of the three women's regiments, and some female pilots moved into, or continued to serve in, men's regiments. The replacement of regimental personnel was standard procedure when its members were injured or killed. Nevertheless, the commander of the all-male Soviet fighter regiment to which Lily Litvyak was reassigned was most reluctant to accept the four women pilots who'd been sent as replacements.[78] Klavdiya Blinova, a fighter pilot from the 586th Regiment, was sent, along with another woman, to reinforce an all-male regiment. She recalls that upon arrival, she and the other woman were not well received, the squadron commanders not at all willing to take "girls." Blinova says the men's comments were "offensive and bitterly disappointing" but they also compelled the women to "break the ice of mistrust" and try to earn their commanders' respect.[79] Blinova notes that the women proved themselves so well that eventually "the squadron became like a family to me and its commander like my father."[80]

When the women in Raskova's flight group began their training at Engels, the flight instructors at the base, accustomed to teaching men, were not happy with the prospect of working with women. They found it necessary to reign in their habit of swearing at the women's mistakes because such responses offended their new students.[81] The flight instructors doubted the women's ability to handle the Pe-2, a new Soviet bomber. Only the women's performance would eventually lay those doubts to rest.[82]

Because it was unusual for women to fly military aircraft, despite their history of service in other branches of the Red Army, their male counterparts were skeptical of the abilities of their female comrades until they were shown that women could perform as well as men under combat conditions.[83] Yevghenia Andreyeva Zhigulenko, a pilot from the Night Bomber Regiment, describes the reaction of a commander at the front when he realized the reinforcements he would receive were female: "Oh god, what are you punishing me for? This is all I need!" It took him a week to make the decision to review his new troops; and another six months elapsed before he changed his mind, telling them, "I was wrong, I repent."[84]

Galina Brok, a flight navigator in the Dive Bomber Regiment, claims that male pilots from nearby "brother" regiments treated the women pilots well. At first the men were doubtful that the "girls" could manage Pe-2 bombers, but they were eventually convinced that their female comrades were equally skilled aviators. Night bomber Marina Pavlovna Chechneva fought alongside a Normandie-Niemen Free French Fighter Regiment in Belorussia, and she notes that the French pilots who flew in Soviet regiments also changed their minds.[85] Galina Turabelidze, a flight navigator, recalls that some French pilots didn't think that a dive bomber could be a "feminine" aircraft. They were amazed to discover one day that the pilots who had just landed a Pe-2 so skillfully during a snowstorm were women, and even more amazed to learn that some of the women flew fighters (Soviet Yaks) exactly like theirs. Years later, these French pilots conveyed their admiration in greetings to all the former Soviet airwomen during a 1960 reunion.[86] Grizodubova said of her experience as the only Soviet woman who commanded a men's regiment, "The men are embarrassed now to remember that in the early days we were not nearly so welcome."[87]

The Germans' perception of women pilots is clear in the following excerpt from a letter Lily Litvyak wrote to her mother while she was still part of the 586th Regiment located at Anisovka Station: "The Germans don't fly here. The girls say, 'they are afraid of us.'"[88] In postwar interviews, German fighter pilots said that in the early days of the war they felt a sense of chivalry toward the female pilots, but that feeling did not last and soon they fought the women exactly the same as men. The women remembered that German soldiers respected courage in battle and would sometimes bury the bodies of their opponents with military honors if they had displayed such valor.[89]

Scant recognition came to the women pilots from the Soviet government. Joseph Stalin's speeches specifically mention the service of women only in their roles as partisan fighters and combat support, but not as aviators. Stalin knew Raskova personally and he had approved the formation of her flight group, so he was certainly aware of their contributions to the war effort, yet never mentioned them publicly. The omission might be explained by his reported response to Raskova when she petitioned to form a group of women pilots: "You understand, future generations will not forgive us for sacrificing young girls."[90]

United States

The American women's experiences depended greatly on the attitude of the commanding officer at their air base. If they simply rode out the initial resistance, they might be able to alter negative first impressions with their performance. But occasionally the resistance to their arrival ran deeper and they could not display their competence because their commander would not let them fly at all. Publicity about the experimental women's program was slight, even within the military itself, so the arrival of women pilots at air bases was often totally unexpected.[91] One woman still remembers her commanding officer's response to her arrival: "I don't like women in general or women pilots in particular. You are not needed here and I will see to it that you don't get into any airplane."[92] Marie Mountain Clark recalls that she rarely encountered resentment from male pilots while stationed at Las Vegas Army Air Base, although she knows that women stationed at other air bases did meet resistance. Clark says that WASPs at Las Vegas AAB had all the rights and privileges accorded an officer in the USAF. The women "rated a salute and saluted all other officers."[93]

A glimpse into public perceptions of the Women's Army Corps during World War II conveys the depth and breadth of the era's gender stereotyping. There were members of the public who assumed that a woman in the military served one purpose only — to provide soldiers with sex. People whose impression of WASP was grounded on that preconception needed more than one exposure to another way of thinking before they could acknowledge that women were equal participants in a nearly exclusively male occupation. Simbeck writes in *Daughter of the Air: The Brief Soaring Life of Cornelia Fort* that, even while wearing uniforms, the women discovered that "Many people were unable to believe [they] flew military airplanes. They were taken for airline stewardesses, members of the Mexican Army, Red Cross volunteers, ferryboat pilots, and even motorcyclists. Being out of uniform didn't always help. On occasion, when they would check into a hotel as a group, people thought they were prostitutes."[94] The persistent image of military women as "home wreckers" dogged all the female pilots, just as it did any other women at the battlefront. Pennington writes of a similar perception about the Soviet women: "The pervasive belief persisted that any woman who went to the front was looking for action in more ways than one. Women who fought could not escape being tainted by a camp-follower image."[95]

Cultural beliefs about females also posed obstacles for the women pilots. Initially, the U.S. Army planned to keep the women grounded during their menstrual periods, but Cochran set out to prove that women could perform equally well at any time. She directed Avenger Field's surgeon, Dr. Nels Monserud, to collect health data on the WASPs, which involved requiring them to report their monthly cycles to see whether their performance was affected during that time.[96] Margaret Ringenberg, 43-W-5, remembers saying "I fell off the roof" to the medical officer, employing the designated code phrase. Ringenberg's account doesn't explain that the reporting was designed to determine whether women could fly as well during their periods as at any other time. She notes that the women were given the option of postponing a flight test for that reason if they wished.[97]

Many trainees were convinced that reporting their periods to the medical officer was mandated to be sure they weren't pregnant.[98] WAFS Florene Miller writes that no one could — or would — verify a woman's report, in any case, so it was impossible to tell if she was lying.[99] Tanner's caption under the photo of Dr. Monserud in her *Zoot-Suits and Parachutes and Wings of Silver Too!* comments on his "strange ideas about women's condition during menstrual periods."[100] Outside the military context, the situation was no different. Although she held a private pilot's certificate, Marie Mountain Clark was denied access to a CAA private pilot advanced course because of the then common belief that the physical stresses produced by acrobatic maneuvers would damage female reproductive organs.[101]

The women were expected to aspire to male competence in the sky, but having achieved it, their employment chances remained the same. In *Sisters in the Sky*, Scharr tells about advising a new pilot:

> "Just fly the best you can. There is no way you can get criticism from men for that. And then when you have shown you know how, what do you think the men will say?"
> "What?"
> "You fly just like a man."
> "That's a compliment?"
> "It's intended for one ... [b]ut it doesn't bring a woman a job in civil aviation."[102]

In *Zoot Suits and Parachutes*, Fagan writes that one male pilot told her that she was, "different from other girls because she could 'talk like a man,'" noting wryly that his words were intended as a compliment.[103] Nevertheless, Fagan was convinced that male pilots generally admired and respected the WASPs as equals.

On one occasion, an expected male reaction failed to materialize. In *Sisters in the Sky*, Scharr describes the night she joined several attractive army nurses on their way to an evening out at an officers' club. However, it was Scharr, without makeup and clad not in a fancy dress but in her uniform, who danced all night with eager pursuit pilots while her mystified companions sat on the sidelines. During their silent ride back to the barracks, the "belle of the ball" kept secret from her companions the fact that she'd flown there in a P-51, a new fighter much coveted by her dancing partners, each of whom had begged her for a chance to fly it.[104]

WASP Elizabeth Strohfus describes giving instrument training to a male friend. As they walked to the airplane, he lagged some distance behind. Finally

Strohfus asked if he was afraid to fly with her and he admitted he was. Strohfus and her reluctant student then agreed that he would perform the takeoff, but she would control the aircraft once it reached 4000 feet. By flying perfect aerobatics at the agreed-upon height, she demonstrated that she could recover from any unusual position in flight, and thereafter, she writes, she had no more trouble "flying instruments" with the men.[105]

—and Rivalries Within

Great Britain

Clearly, the women fought the preconceptions of others, but things were not always harmonious even within their ranks. The dividing line between upper and lower social classes in England persisted into the ATA, and they were especially prominent between the original and subsequent members of the Women's Section. Differences in social standing formed the widest gulf among them. Amy Johnson, who had set a record in 1930 as the first woman to fly alone from England to Australia, was miffed that her friend Pauline Gower had been chosen as leader of ATA's new Women's Section, despite the fact that Johnson thought she was more qualified for the position due to her fame for breaking aviation records. Although she felt entitled, Johnson's bid to join Britain's Civil Air Guard was apparently not taken seriously by the CAG. It rejected her as a member because she had built a reputation for being as interested in earning money as in achieving new goals in aviation and it was assumed that she would not be interested in the position after she learned that it was unpaid. If Johnson had ever been considered to lead the Women's Section, it is likely that Gower prevailed on the basis of her standing in British society, not to mention her skill as a pilot and her success as a businesswoman. According to the ATA's leader, Gerard d'Erlanger, Gower was "One of Us," while Johnson, a woman born into humbler circumstances, clearly was not.[106]

Soviet Union

In the Soviet Union, the 586th Fighter Aviation Regiment appears to have had significant personnel issues, according to the recollections of several of its members who attribute those difficulties to the personality of their commander, who was engaged in a bitter competition with Raskova for supremacy. Pennington's *Wings, Women and War*, a history of the three women's regiments, tells what effect their leaders' personalities had on each regiment's ultimate success.

The first commander of the 586th Fighter Aviation Regiment, a pilot appointed by Raskova, was Major Tamara Aleksandrovna Kazarinova. Kazarinova had been wounded during an air raid and she walked with a limp. She often worked late into the night on battle plans. However, despite her obvious dedication to her regiment's success, she was dismissed after serving as commander for only six months.[107] The ostensible reason for Kazarinova's removal was that the injury to her leg had worsened and because of it, she was unable to fly the fighters her pilots used in battle.[108] Some accounts claim that Raskova and Kazarinova were rivals caught in a bureaucratic tangle of the Stalin era. A few women of the 586th Regiment recall that Kazarinova's manner was stern and she was widely disliked. Some thought she showed poor judgment as a commander and fostered discontent among her pilots, some of whom tried to remove her from command. One of those pilots, Lidya Litvyak, was among the eight women transferred from the 586th to the front at Stalingrad to replace pilots who'd been killed. All of the women that Kazarinova transferred out of the 586th had been vocal opponents. Of the eight, four were dead by 1943.[109]

Questionable circumstances also surrounded the death of pilot Valeriia Khomiakova, a former member of an aerobatic team and the first woman to shoot down an enemy aircraft at night.[110] Having traveled the long distance to Moscow to be recognized for that feat, the fatigued Khomiakova, upon her return to the regiment, was immediately assigned night alert duty. Kazarinova ordered a mechanic to warm up an airplane while Khomiakova slept in the dugout until the airplane was ready for takeoff. The suddenly awakened pilot jumped into the plane, but without the benefit of lights to guide her takeoff, she crashed into an obstacle and was killed. According to the regiment's next commander, her vision wasn't adapted to the darkness. The pilot's death was attributed to a combat loss, and as such, it was never investigated. It was this incident that brought about Kazarinova's dismissal in October 1942.[111]

The replacement for Kazarinova was Major Aleksandr Gridnev, his assignment to a women's regiment most likely constituting a penalty, because Gridnev had been arrested once as an "enemy of the people." The women's squadron that had been transferred to Stalingrad was replaced by a men's squadron, also probably because they were being punished for some infraction, a common tactic of the Stalin era. The women of the 586th would later comment that they considered Gridnev a good commander.[112]

United States

In the United States, Nancy Love's WAFS and Jackie Cochran's WFTD never really melded into a cohesive unit, their attitudes reflecting the competition already thriving between their two leaders. The nearly concurrent formation of two groups of women pilots led to tensions that persisted long after Love's group was folded into Cochran's to form the WASP. Even when they were posted together at American air bases, the two groups often appeared prejudiced toward each other. The memoirs of women who began their service under Love and those who were recruited by Cochran vary markedly in their opinions about the merger and their perceptions of why Cochran became leader of the whole organization. Some women claim that Cochran's wealth gave her an edge and made it inevitable that she would prevail. Others are convinced that Cochran's forceful personality ensured her the top position because she simply bullied her way into command.

Love's and Cochran's rivalry took root with the nearly simultaneous beginnings of the WAFS and WFTD. They held different views about the role women pilots could serve for the United States military and thus they did not agree about the minimum qualifications required for entry, nor did they see eye to eye about the extent of training the volunteers would require. Byrd Granger, who entered in the first WFTD class, 43-W-1, wrote a detailed history of both groups in her book, *On Final Approach: The Women Airforce Service Pilots of W.W. II*. Granger takes a broader perspective and claims that there was a place for women pilots in every aspect of wartime flying. She writes that Love's plan for using women pilots — recruit highly experienced pilots who were ready to perform — and Cochran's — recruit pilots with little experience but provide them extensive training — were equally valid and need not have been competitive.[113] Granger was among the twenty-eight women who were "guinea pigs," the first trainees in Cochran's program. As editor of the *Fifinella Gazette*, Granger was perhaps more aware than most about happenings in the unusual little flight school.[114] The new program's growing pains included misunderstandings concerning the women pilots' ultimate role and their relationship to the already established group of WAFS, feelings that intensified as more trainees arrived in Houston.

When Cochran learned that pilots from 43-W-1 had been assigned ferrying duties at air bases selected by Love, she lobbied for greater control over her training program. She wanted her pilots to perform all kinds of flying tasks, not just ferrying. In time, she was appointed director of women pilots flying for the Army Air Forces and was also appointed assistant to the chief of the AAF staff in Washington.[115] Rickman's biography of Nancy Love hints that Love vented her frustrations with Cochran in letters to her husband and she reportedly burned her WASP uniform as soon as the program disbanded. Nevertheless, exaggerations about the two women's rivalry are rampant. One reason could be that rank and file pilots never received that much information.

4—*Becoming Military Pilots* 75

In winter 1943, WAFS posed in their official uniform overcoats and overseas caps in Wilmington, Delaware. Shown are (left to right) Helen Mary Clark, Nancy Batson Crews, Helen McGilvery, Teresa D. James, Gertrude Meserve Tubbs LeValley, Esther Nelson Carpenter, Betty Gillies Huyler, and Dorothy Fulton Slinn (The Woman's Collection, Texas Woman's University).

As Kay Gott, 43-W-2, notes in her *Women in Pursuit: Flying Fighters for the Air Transport Command Ferrying Division during World War II, A Collection and Recollection*, the women simply didn't know the reasons behind their commanders' decisions and that ignorance spawned myths that circulated widely. One former WASP comments: "I didn't appreciate Cochran. I didn't realize that it was her idea right from the start that woman [*sic*] should be given transition up to their capabilities; and General Arnold backed her up. You see, I never appreciated her. I never knew that this was one of the conflicts she had with the Ferrying Service."[116] As subordinates, the women could only speculate. For example, Clark remembers having only *one* female instructor for advanced flight training, a woman named Helen Duffy; but after only one flight, she never saw Duffy again, nor was she told the reason why.[117]

Advocates for one leader were prepared to assume that nefarious motives lay behind the actions of her counterpart. For example, Cochran was the first American woman to fly a Lockheed Hudson bomber to England, just before the United States entered the war, although she was not allowed to perform either takeoff or landing.[118] The flight occurred before Japan attacked Pearl Harbor, an important factor in understanding why she received permission at all. Sometime later, Love and her friend, WAFS Betty Gillies, in a bid to assist with a large delivery of bombers to England in August 1943, prepared to fly a B-17 bomber across the Atlantic. Bad weather delayed them for twenty-four hours, so the two women were still on the ground in Labrador, Canada, when orders came from General Arnold to cease their flight immediately.

Arnold, who had allowed women pilots to fly for the Army Air Force, providing they flew within the boundaries of the United States, flatly refused to allow American women to enter a war zone. Because Arnold's reason for aborting the flight was not generally known, rumors took root among Love's supporters that Cochran had interceded with Arnold to end Love's and Gillies's flight out of jealousy. In *Sisters in the Sky* Schaar speculates that it was Cochran who foiled Love's transatlantic flight.[119] She isn't alone in her assessment. Nancy Bird, an Australian commercial pilot during the interwar years, writes in *My God! It's a Woman* that no reason was ever given for Arnold's order. Bird then restates the story that Love and Gillies were *on the runway* when the command came from Arnold to "cease and desist," a myth that is contradicted by other sources and later by Love and Gillies themselves. Bird, having met Cochran, thought the WASP commander rude. She further speculates that Cochran engineered the WASP disbandment to avoid having to serve under WAC Colonel Hobby, claiming that Cochran ensured that no woman should fly military aircraft in the United States thereafter, a "decree" lasting thirty years.[120]

Some women assume that the assertive Cochran pushed her way into the WASP directorship.[121] WAFS Evelyn Sharp writes in July 1943 that Cochran's wealth gained her the exclusive command of the WASP.[122] According to Granger, Cochran tended to fuel the fires of her rivalry with Love, especially when she referred to the WAFS as "Nothing of importance. A bunch of society dames."[123]

5

Daily Work in England and America

For the most part, British and American women were restricted to domestic flights only, and they neither trained for nor flew in combat. Their work was designed to support male pilots who did. In contrast, the volunteers for Marina Raskova's flight group became part of the Soviet Union's Red Army and they performed whatever there was to be done, including flying fighters and bombers in combat alongside male pilots. That single distinction clearly separates the Soviet women's daily work from the service performed by the women of England and America. Therefore the Soviet women pilots' daily work must be considered separately. Their experience is part of the following chapter on hazards and sacrifices.

Assignment to Air Bases

Great Britain

The first eight ATA women were all sent to a single ferry pool, Hatfield, because there were so few of them. As the Women's Section grew, a second women's ferry pool opened, and seven of the women were transferred to Hamble, under the command of Margo Gore. In time, Hamble became the only all-female ferry pool in England.[1] Curtis notes that Hatfield, where she had been stationed, was closed in 1941 and the eleven women still serving there were moved to Hamble.[2] As the ATA organization grew, the flight schools' enrollment also expanded and sometimes trainees were relocated. Women were moved to other ferry pools that had previously been all male. Diana Barnato Walker joined the ATA late in 1941. She remembers that by then the women were given two-week temporary placements at a ferry pool to see how things would work out before receiving a permanent assignment.[3] Walker was stationed at White Waltham, located close enough to her family home that she could continue to live there. The American, Hazel Raines, wrote letters home during her service which were later compiled by her niece, Regina Hawkins, in *Hazel Jane Raines: Pioneer Lady of Flight*. Raines' letters provide an entertaining description of her stay at the home of Lady Astor in 1942 while serving with the ATA.[4]

United States

The small group of original WAFS gathered at New Castle AFB in Wilmington, Delaware, where Nancy Love maintained an office.[5] After Cochran's WFTD program began producing comparatively larger numbers of pilots, the WAFS relocated to air bases with established ferrying units. A group of WAFS settled into Long Beach, California, where the 6th Ferrying Group was located. Besides Long Beach and Wilmington, there were women's ferrying groups stationed in Romulus, Michigan, and at Love Field in Dallas, Texas.[6]

The training program for Cochran's WFTD began in Houston, Texas, with the arrival of the twenty-eight women who made up its first class. As subsequent classes arrived each month, it didn't take long for the new program to outgrow its limited space, so the training program moved to Avenger Field in Sweetwater. The women who had completed WASP training were stationed at more than one hundred airbases throughout the United States. Early in the program, they were allowed to choose where in the country they preferred to be; but later the rules for them were tightened in accordance with air force policy, and they were purposely assigned to air bases that were not located near their homes.[7] Margaret Ringenberg, 43-W-5, chose to be stationed at New Castle Air Base in Wilmington.[8] Marie Mountain Clark, 44-W-1, requested the southwestern United States and, after the ten days of leave that followed her graduation, she reported to Las Vegas (Nevada) AAB, one of the first WASPs to arrive there.[9] O. Vivian Fagan, 44-W-7, also chose the general location of her air base after she graduated, and she was stationed at the Stockton (California) Army Airforce Base.[10] The official WASP website lists a total of 120 air bases where the women served. They range, in alphabetical order, from Alamogordo Army Air Base in Alamogordo, New Mexico to Yuma Army Air Field in Yuma, Arizona.[11]

With their training complete, the graduates sometimes found themselves quite isolated at their bases, a factor that impacted their sense of connection to their WASP cohorts. Ann Baumgartner Carl, 43-W-5, worked as a test pilot at Wright Air Force Base in Dayton, Ohio, often the only woman pilot working there, so she rarely saw her contemporaries, and therefore did not experience a sense of connection to her fellow WASPs until she attended reunions years later.[12] Sometimes their isolation only increased their other difficulties. Florene Miller Watson, remembering her early days as a female pilot, describes "flying in a man's world" prior to joining the WAFS, and she recalls that, because of that, women pilots tended to form an immediate bond.[13]

Accommodating the housing needs of female personnel was challenging on air bases previously used exclusively by men. Sometimes the American women were housed in nurses' quarters, sometimes in bachelor officer quarters (BOQ).[14] The women who performed combat training tasks such as towing targets, or those who worked as test pilots, stayed on their air bases every day. But the pilots who ferried airplanes could potentially find themselves almost anywhere in the country at day's end and they sometimes had to "remain over night" (RON) where they'd landed because they were not supposed to fly after dark. Occasionally they needed to wait out unfavorable weather conditions. If those airfields lacked separate rooms for them, they tried to find rooms in nearby hotels.

The Roles of Gower, Love, and Cochran

After conducting personal interviews, Pauline Gower chose the eight women who were to be the first female pilots in the Air Transport Auxiliary forming the core of the Women's Section. Gower's office was originally located at Hatfield, home of the de Havilland Company, which manufactured the light training aircraft used by the RAF, but while she was stationed there she resided in a private home.[15] One of her first responsibilities as a leader was countering negative public opinion about letting females fly RAF airplanes. To that end, she wrote articles and participated in radio interviews in an attempt to allay fears that the women would destroy expensive government property while simultaneously taking valuable work away from men.

Gower also worked to expand the number of pilots in the Women's Section. As more airplanes were manufactured, demand increased for pilots to ferry them. Soon after recruiting the first eight women, Gower was allowed to choose two more, one of whom was her friend from her days at London's Stag Lane flying club, the fabulously famous Amy Johnson. Although she held many aviation records, Johnson was not confident in her ability to pass an ATA flight test and she was further unsettled by the presence of other candidates who were obviously awed by her celebrity. Gower coaxed her friend to undergo the test and reassured her that, although required, it was more or less a formality in Johnson's case. Johnson passed her flight test and entered ATA as a second officer on May 20, 1940.[16] Gower recommended adding even more pilots to the ranks of the Women's Section, her quiet persistence and diplomatic skills gaining her authorization to boost their numbers by an additional ten women.

Several of the pilots under her command remember Gower as kindhearted and genuinely interested in their welfare and well-liked by the women. Her sense of humor and fun combined with high standards and an inclination toward absolute fairness. Gower knew that her pilots' performance had to meet the highest standards. In dealing with personnel issues, she often displayed a lighthearted personality, but she could be deadly serious about safety issues involving the women, especially where the aircraft were concerned. In this regard, she upheld the ATA's requirement that accidents were to be avoided at all costs. Gower's duties also included tasks associated with fatalities in the Women's Section, the first instance being the disappearance and death of Amy Johnson, the first woman killed serving in the ATA. After it was certain that Johnson's airplane had been lost somewhere over the Thames estuary in frigid fog while on her way to deliver it in January 1941, Gower notified Johnson's parents and then wrote her friend's obituary.[17]

Gower's office was moved from Hatfield to the ATA headquarters at White Waltham in November 1941 because it was a better location for administering the expanding Women's Section. There, too, she resided away from her air base. Gower took her turns flying the taxi aircraft that carried ATA pilots back to their home bases, and she visited the ferry pools where her pilots were stationed. She gave flight tests and checked aircraft. According to Fahie, her son and biographer, in 1940 his mother visited sixteen RAF stations and made sixty-eight air-taxi trips, bringing pilots back to their air bases after they had made deliveries.[18] Although she seized any opportunity available to fly, her administrative duties kept her grounded after she relocated to White Waltham.

Gower received guests to the Women's Section. England's King George VI and Queen Elizabeth visited the White Waltham headquarters in 1942.[19] That same year, she entertained another distinguished visitor, Mrs. Eleanor Roosevelt, First Lady of the United States, who visited the Women's Section in October.[20]

By 1943, when pilots of the Women's Section were flying virtually all types of operational aircraft used by the RAF, they continued to receive only 80 percent of the pay rate earned by male pilots. In an era when equality in compensation was virtually unknown, Gower and a few others lobbied for a change to the British treasury ruling, and, as a result of their efforts, by June 1943, an ATA woman earned the same as an ATA man.[21]

A major personal obstacle impacted Gower's ability to carry out her duties. Her health had been problematic since suffering a serious illness in childhood, and at times she had to limit her activities when her doctor ordered her to rest. Fahie speculates that his mother would probably not have passed the ATA's physical exam, especially in later years when its requirements became more stringent.[22]

Nancy Love's first job as leader of the WAFS was conducting interviews of pilots for an anticipated 90-day commitment to the ferry command. She selected the first WAFS and arranged for her new recruits to be housed in bachelor officer quarters 14 on the Newcastle Army Air Base in Wilmington, Delaware. Love appointed her friend Betty Gillies as second in command of the WAFS at Wilmington. Gillies served as their resident housemother, but later Love appointed another woman, a Mrs. Anderson, to watch over the "girls," despite the fact that several of them were married and a few had children.[23] As the women's ferrying squadron grew, Love chose individual WAFS for leadership roles wherever separate contingents of female ferry pilots were stationed.

Love spent much of her time learning to fly as many types of military aircraft as possible. Her plan was that she, a woman of average size and strength, would gauge her own ability to manage an aircraft successfully before allowing her WAFS to do the same. Love also travelled to the several air bases where her ferry pilots were stationed. Although she transported some airplanes herself, her ferrying work was curtailed by the demands of her administrative duties.[24] Nancy Love and her WAFS were subordinate to Jacqueline Cochran, the director of (all) women pilots, and Love's biographer speculates that one motivation for her push to transition WAFS to more sophisticated aircraft models was her desire to keep her pilots always one step ahead of Cochran's. Love continued adding pilots to her ferrying group until General Arnold informed her that the WAFS could not employ any woman who was *not* trained in the WFTD. Lenore McElroy, whom Love hired just prior to Arnold's ruling, thus became the last member of the original WAFS.[25]

In her capacity as head of the squadron of women in the army's ferrying division, Love attended to the necessary business surrounding the deaths of women under her command. Cornelia Fort was the first of the WAFS to be killed in service when her airplane collided with another in the skies over Texas. Rickman writes that Love could not bring herself to speak at Fort's funeral, but she attended the service along with WAFS Barbara Erickson. When two more women were killed, Love did not attend their funerals, sending a representative instead.[26]

Jacqueline Cochran interviewed Roberta Sandoz Leveaux in Fresno, California, the first step in Leveaux's journey to Montreal to prepare for service with the group of pilots who were going to fly with the ATA. Leveaux remembers that Cochran stressed the importance of two things during that interview.

Nancy Love at work in September 1942. Love founded the Women's Auxiliary Ferrying Squadron and "transitioned" to as many different airplane types as possible to prepare the way for her WAFS to do the same (courtesy Library of Congress).

First, it was vitally important for the women to perform well in England because their success could make a similar program possible in America. Second, the women should hold no illusions about the work they would perform — ferrying airplanes would be difficult and decidedly unglamorous.[27]

When she returned from the success of her venture with the ATA, Cochran was already convinced that a larger — though less experienced — group of women pilots could be trained for many different kinds of flying in the Army Air Force and she designed her training program to accomplish the goal of versatility. Cochran used her own airplane to check on the women under her command. She had learned to fly a twin-engine bomber in 1941 before the WASP program began, but she saw no need to personally transition to each aircraft type that her pilots would fly as long as they themselves were able to advance to more sophisticated aircraft. Cochran apparently tried to recruit likely candidates to instruct in her own training program. She visited Phoebe Omlie's Tennessee flight school in January 1943, hoping to entice the school's ten female flight instructors to join her pilot training program, but the women declined Cochran's offer because they were committed to finishing Omlie's course.[28]

As the director of women pilots, Cochran undertook all the administrative duties which are recorded in her *Final Report on Women Pilot Program*.[29] During the war years, Cochran's flying was confined to trips from her command center office in Houston to the many air bases where she was needed, either as an observer or as a mediator. Appropriate to her role as their director, Cochran attended the funeral of Mabel Rawlinson 43-W-3, a WASP who had been killed at Camp Davis in North Carolina.[30] Cochran participated in WASP graduations, and she approved the design of the graduates' silver wings, paying for them from her own pocket. If the family of a WASP who had died either in training or in service could not afford to send for their daughter's body, she covered those expenses too.

Aircraft Conversion and Transition

The women learned to fly airplanes that were larger and more advanced than simple trainers through the efforts of their leaders, who wanted their pilots to fly all the military aircraft currently in use. At first, women pilots could fly only basic trainers, lightweight and relatively simple aircraft. They had taken their flight tests in these trainers and no one in any air force was willing to trust them with more complex models. The process of advancing from training aircraft to more sophisticated models was called "conversion" in England and "transition" in America.

The sole function of Britain's Air Transport Auxiliary can be summed up in the numerous interpretations of its initials, ATA. Rosemary Rees du Cros claims the letters stood for "Any aircraft, To any address, At any time," or, alternatively, "Ancient Tattered Airmen," a reference to the aging pilots from World War I serving their country in yet another world war.[31] Fahie titles one chapter in his biography of Gower "Always Terrified Airwomen," providing another variation on the theme.[32] Hazel Raines cites the organization's slogan as "any Aircraft, any Time, any Address."[33] It can be reduced further still. Render, in *No Place for a Lady*, states the abbreviation thus: "Anything To Anywhere."[34]

Given that charge, it was imperative that ATA pilots be as versatile as possible, and, although Gower herself would never learn to fly larger aircraft, she made it possible for her pilots to do so by attending a flying school which had never admitted women before. Restrictions on the aircraft types flown by the Women's Section relaxed gradually, not only as the women proved their competency, but also because their commander insisted that they be allowed to convert to the advanced models.[35]

The first eight women flew only slow open-cockpit trainers because the RAF was unwilling to trust its valuable fighters to females. Gower knew that restricting her pilots to the lightest and simplest airplanes would hamper their usefulness to the organization, so she convinced officers at the RAF's Central Flying School to admit women, posing her request as an opportunity to take on the *challenge* of training females to fly more powerful aircraft.[36] Conversion was a crucial stepping-stone if the women were ever to perform the full range of ferrying tasks; and for the women, it was accomplished in five stages. British aircraft models fell into six classes: Class I, single-engine light; Class II, single-engine service; Class III, twin-engine light; Class IV, twin-engine service; Class V, four-engine; Class VI, flying boats.[37] Learning to fly heavy four-engine bombers was the final step in the process. Clearing that hurdle gave women the opportunity to fly all classes of aircraft used by the RAF with the exception of "flying boats," from which they continued to be restricted, not due to their lack of capacity, but on *moral* grounds. The rationale was that flying boats were never flown solo and a mixed-sex crew presented potential problems in the event of a forced landing. Fahie observes that Jackie Cochran's trans–Atlantic bomber flight to England in 1941 also may have helped further the cause of ATA women training on advanced aircraft types.[38]

By the time Diana Barnato Walker joined the ATA, women were already being trained on all five classes of British aircraft. Walker points out that once a pilot had mastered an airplane in one category, there would be little difficulty in flying the others in that same class. Because their instruction program began with the lowest level, Class I (trainer) aircraft, later trainees were never able to progress beyond Class III (light twin engine) before the war ended.[39] Fahie describes a four-part conversion to heavier aircraft beginning with fighters, then twin-engine bombers, and ending with heavy bombers. Between these training periods, the women had an opportunity to ferry the new aircraft type so they could become proficient in it.[40]

Although they had learned to fly different kinds of aircraft, it was to the ferry pilots' advantage never to become too familiar with a single type because that familiarity could impede their ability to make a rapid transition to other types, a necessary skill for ferry pilots who were expected to transport many different airplanes. Rosemary Rees du Cros claims that all the ATA pilots "got extremely childish about collecting 'types,'" and states that she herself had flown ninety-one different types in all, but developed her greatest proficiency with the taxi airplane, the Avro Anson.[41]

While she was in ATA, Leveaux learned to fly fifty-four different aircraft types.[42] Despite the restrictions surrounding flying boats, du Cros describes her wartime experience with the lumbering seaplane, the Walrus, in her memoir, *ATA Girl*, and she goes on to say that the American Airacobra was unpopular among the ferry pilots because of its long flimsy-looking nosewheel and the fact that the pilot sat astride the crankshaft of the engine, which was located behind the seat. Du Cros claims that was why the airplanes were quickly sent on to the Soviet Union, by then a participant in America's Lend-Lease program, and in her memoir she muses about how many *Russians* the Airacobra may have killed![43]

Transition in the United States followed much the same pattern. At Avenger Field the women advanced from basic training aircraft to heavy four-engine models. Throughout their service, they trained at other locations too, so they could acquire proficiency in both pursuits and bombers, and even some early jet aircraft.

Daily Work, Daily Life

Great Britain, and Cross-Country Ferrying in Britain

After they were assigned to ferry pools, the British women lived in private homes or maintained their own apartments, there being no accommodations for them at places which had previously been all-male air fields. In *New Wings for Women*, Helen Harrison describes her own ATA routine: "We get up about 8 A.M., on our jobs here, have breakfast, and take a bus to the airdrome. Then we pick up our 'chits,' or assignments. Then we check on the weather, find a taxi plane to transport us to some plane factory where we are to collect a plane for delivery."[44]

Daily life was never precisely that predictable because a ferry pilot's work always depended on the vagaries of the island nation's weather. If the weather was favorable, they took to the air and stayed there as long as possible before darkness fell. An operations officer at each ferry base set the pilots' daily schedules, planning delivery routes that would make the most efficient use of available human and material resources. In her memoir, du Cros describes the logistical responsibilities of the person who arranged the daily flights.[45] Milstead also outlines the considerations necessary for making ferry assignments in Render's *No Place for a Lady*. The operations officers were to move aircraft with as few stops as possible, and by the shortest route in order to use the least possible fuel; to return all pilots to their bases at the end of the day; and to evenly distribute the work among the pilots according to their capabilites.[46] Every morning, when the weather allowed, a pilot received her ferry chit, the authorization to pick up and deliver an airplane. When bad weather made flying impossible, the women waited in the mess playing cards, knitting, talking or sleeping. Because their schedules alternated between long days of flying and equally long days of waiting, most pilots learned to catch catnaps whenever they could. Leveaux, in her account, *Americans in the British ATA*, recalls her routine of catching the nine o'clock news, a ritual when she served with the Women's Section.[47] The British women were granted weekend-long leaves during their service, but for the Americans, who had signed on for eighteen months of service, and for others whose homes were far from the British Isles, leaves were longer, allowing enough time to travel home.[48] The American Hazel Raines, for instance, received a two-month leave in September 1943.[49]

The ATA moved airplanes from factories to maintenance units where radios and guns were installed in preparation for flying them in combat. A primary responsibility of the ATA was moving new airplanes away from the factories where they were manufactured to where they would be fitted out for battle with the addition of radios and weapons. Factories made desirable enemy targets, especially when aircraft were left there, so their work was critical. By performing this simple task, the ATA freed Britain's fighter pilots to be used where they were most needed — in combat.

If the planes needed repairs, the ferry pilots flew them back to maintenance units.[50] Walker remembers that the ATA also delivered war-weary aircraft to their last landing places.[51]

During the long summer days in England's northern latitude, the pilots flew many hours. Conversely, in winter, they flew far fewer hours. As long as deliveries went as planned, the women hitched rides back to their point of origin from an Avro Anson taxi airplane. But when it was too late in the day or they were too far from the ATA air taxi service, they might be stuck at airfields with no accommodation for females so they had to hunt for a place to stay overnight after a long day of flying. Occasionally they were rerouted to make an additional delivery. Leveaux, who was assigned to the no. 15 ferry pool at Hamble under commanding officer Margot Wyndham-Gore, recalls that she, along with other uniformed ATA personnel, was often housed in unused rooms in private homes.[52] The pilots sometimes caught a train to return to their home ferry pool. On night trains they might snag a sleeper car if they were lucky. Veronica Volkersz describes her work flying Tiger Moths to Scotland: "For the next three or four days the routine was similar: up to Prestwick during the day, then back home on the night train. Although we managed to obtain sleepers every time, it was a tiring business. When we progressed to the more advanced aircraft we often got a return ferry trip south, but while we were restricted to light aircraft we had to get back by train."[53]

In cross-country ferrying in Britain ATA pilots were forbidden to "fly blind," indeed they received no instruction on flying with instruments. The airplanes they transported had not yet had radios installed for security reasons and so as not to interrupt other radio transmissions. ATA pilots flew only in daylight and in what their pilots deemed acceptable weather. Although they carried maps, the pilots were not allowed to mark them, in case the maps should fall into enemy hands. Sometimes pilots had no firsthand knowledge of an airplane they were assigned to fly so they relied on "Ferry Pilots Notes."

Roberta Sandoz Leveaux remembers the reassurance her set of Ferry Pilots Notes gave her. Leveaux also recalls that there were highly detailed descriptions of all the aircraft flown by the ATA available at each ferry pool. She writes that an unprepared pilot had only herself to blame because the information was always at hand.[54] Lettice Curtis, the ATA's historian, describes Ferry Pilots Notes as 4" × 6" cards with all the essential information about each airplane flown condensed on both sides of a single card and updated as needed. The cards were perforated and held together by rings so a pilot could conveniently carry in a pocket notes for every aircraft in use.[55]

Milstead acknowledges that the flying done by ferry pilots and by combat pilots in the same airplanes was very different. She says that transporting an airplane from point A to point B was much less demanding than flying that same airplane in combat, despite the challenges of weather and other difficulties. Milstead praises Ferry Pilots Notes for simplifying the transport of so many different models of aircraft. She relates from her own experience the following exchange:

> "Have you got much time on this type?"
> "No, I've never flown one before."
> "Never flown one before?! Then what makes you think you can handle it?"
> "No problem. It's all in my book."
> "In your book! Good God, girl, you can't fly this aeroplane from a book!"
> "I can — from *my* book."[56]

ATA women were forbidden to fly across the English Channel during most of the war. Until the very last days of fighting they flew only in the British Isles. As the tide of war turned, the pilots of ATA made deliveries on the European continent, including shipments of supplies to Belgium in late 1944. Diana Barnato Walker obtained permission to accompany her RAF husband on one delivery mission during a leave from her ATA duties. Walker later speculated that the officials who cleared her cross–Channel flight may not have figured out that she was female because her husband wrote only her initials on the request rather than her full name, though in another line the prospective pilot is referred to as "she."[57] Walker's flight to Belgium in formation with her husband's Spitfire was uneventful, and afterwards the couple remained on the Continent for a time. On their return, the two were delayed by fog, but eventually it cleared and they took off. Flying back across the Channel in a Spitfire unequipped with a radio, Walker noticed the weather closing in again. Having lost sight of her husband's airplane, Walker recalls that she almost accidentally "found" England in the midst of the fog. The couple's trip was featured in the *Daily Mail* on 17 November 1944. After Walker had completed what happened to be the first flight by an ATA woman into Belgium, other members of the Women's Section were given permission to fly to the Continent and some of them began making routine deliveries of aircraft to the newly liberated country.[58]

According to Rosemary Rees du Cros, four ATA women trained as flight engineers, and each one served as the lone crew member accompanying the pilots of four-engined bombers. Their role was to check the aircraft during its flight and again upon landing.[59]

United States

Cross-country ferrying was the first kind of work performed by American women pilots. They moved airplanes from factories to ports where they would be partially dismantled and shipped overseas. At the other end of a warplane's journey, it might reappear in the United States in "battle weary" condition and be flown by a WASP to its ultimate destination — a scrapyard. If damaged airplanes were deemed salvageable, test pilots, including WASPs, determined their airworthiness before they returned to battle. Advances in technology during the Second World War brought new equipment into use, which was tested by WASP. Some of the WASPs learned to fly the most modern aircraft of the day, including jets.

In the United States, WASPs performed tasks consistent with their status as utility pilots whose role was to free other pilots for combat. WASPs trained ground artillery troops by towing fabric targets a few hundred feet behind an airplane, their first assigned responsibility in addition to ferrying. Sometimes they made shallow dives with airplanes so trainees on the ground could practice their shooting, at times using live ammunition. They engaged in tracking and searchlight missions. By late 1943, WASPs participated in simulated strafing, smoke laying and other chemical missions, radio control flying, basic and instrument instruction, engineering test flying, and administrative and utility flying.[60] The WASPs also served as couriers and they transported military personnel. Many women in the United States had worked as flight instructors, thanks to the Civilian Pilot Training Program, so they were able to teach others how to fly and how to rely on instruments in flight with Link trainers (ground simulators for teaching instrument flying). As the ATA did, the WASP played an important role in moving airplanes from factories to airfields. The ferry pilots never knew which type of aircraft they might have to fly, so they needed to be conversant with a wide variety of types.

When the scope of America's Lend-Lease program expanded, aircraft were ferried from factories near Niagara Falls in the United States for delivery to Soviet pilots who had come through Siberia into the U.S. territory of Alaska. When they flew this route, the WASPs stopped to refuel their airplanes at East Base in Great Falls, Montana. Then the women would turn their airplanes over to male pilots who flew them through Canada to Ladd Field in Fairbanks, Alaska. Continuing on into Alaska was considered too hazardous for females. When the women asked why they couldn't fly the airplanes they'd just landed, if the men could, they heard the explanation that no facilities for women existed at the Alaska base.[61] The book *Hazel Ah Ying Lee*, by WASP Kay Gott, who flew some of the Lend Lease airplanes into Montana, shows a photo of three rows of Bell Kingcobra P-63s, a total of seventy-two aircraft awaiting transport on the runway at the Bell Factory in Niagara, New York.[62] They carried extra fuel tanks on their wings for the long flight from Fairbanks to the Siberian refueling station. They are painted olive drab and display the Russian red star outlined in white. Gott quotes General William Tunner of the Ferrying Division, who told what happened to those P-63s when their power was cut: they assumed the "glide angle of a brick."[63] In winter months, weather delays along the northern delivery route were common.

WASPs delivered other new aircraft from factories to their ports of embarkation; they ferried training planes to air bases and liaison craft to army bases. According to Strohfus, in 1943 and 1944 women performed 90 percent of the aircraft ferrying done in the United States and 85 percent of all the flight training.[64]

Some of the WAFS also added administrative tasks to their work if they were chosen as leaders of the women stationed at ferrying bases. Barbara Erickson led the group of WAFS at Long Beach, California. Adela Scharr supervised the group stationed at Romulus, Michigan, and Florene Miller was in charge of the WAFS at Love Field in Dallas, Texas.[65] Helen Richey was tapped by Jacqueline Cochran to replace her in England and assume charge of the American women still working in the ATA after Cochran returned to America.[66] Richey was dismissed from the ATA, however, not long after Cochran departed. It seems likely that she was let go for wrecking too many airplanes. Kerfoot, her biographer, speculates that Richey found the burden of the dual role of ferry pilot and group commander too taxing, especially when it was combined with concern about her mother's illness back home, so she resigned, making her last flight over England on January 9, 1943.[67] However Whittell writes that Richey's accident, which damaged a twin-engine Vickers Wellington in January 1943, left Richey able to walk from the wreck smoking a cigarette but got her dismissed.[68] According to Whittell, Richey accepted Cochran's standing offer to the American women who flew with the ATA — they would have a "place among the WASPs."[69] Taking Cochran up on her offer, Hazel Raines returned to the United States in 1944, after fulfilling her term with the ATA, and then she trained in class 44-W-3 and flew with the WASP.

WASP Betty Greene, in *Flying High: The Amazing Story of Betty Greene and the Early Years of Mission Aviation Fellowship*, writes that Cochran tapped twenty-five pilots from class 43-W-3 for new kinds of work at Camp Davis.[70] In her memoir, *Women Who Dared*, Yvonne Pateman, 43-W-5, states that the group's selection was an experiment to determine whether women could fly effectively and successfully in a combat-oriented environment.[71] Greene worked in a tow target squadron at Camp Davis. She and her classmate, Ann Baumgartner Carl, tested artillery tracking by flying low in small L-5s for hours. They also flew at 10,000 feet for radar tracking practice in an A-24 dive bomber, logging many hours in the air on the hard cockpit seats. Along with male pilots, they towed targets in a Lockheed B-34 and flew searchlight

5—*Daily Work in England and America* 91

tracking missions at night. They also flew remote-controlled drones from twin-engine C-78s. Between periods of assigned work at Camp Davis, they took an occasional cross-country flight or practiced flying in formation.[72] Greene enjoyed her time at the airbase because of the variety of flying she was able to do there.

Greene and Carl were transferred to Wright Field Air Force Base in Dayton, Ohio, where both women were assigned to the equipment lab and the stratosphere research project. They tested oxygen masks and electric flying suits at high altitude and low temperature in Wright's stratosphere research project.[73] The reason for stratospheric testing was to determine whether American pilots could attack German planes at high altitude without suffering embolisms from the rapid decompression following high altitude flying.[74] Although WASPs did not directly participate in these activities, they watched experimental high-altitude parachute jumps. One of these tests resulted in the death of a man attempting to jump from 40,000 feet. His instructions were to free fall to an altitude of 20,000 feet before opening the parachute, but he apparently lost consciousness before reaching the lower altitude.[75]

Carl was the only female test pilot at the Wright Army Air Base in Ohio. Other women posted there were involved in different kinds of flight testing, and although she and Betty Greene had been assigned to Wright together, Greene stayed in the equipment lab. Carl was assigned to the Fighter Flight Test branch, where she flew the latest experimental aircraft, including the first jet-powered airplane, the XP-59A, the first American woman to fly a jet.[76]

Adela Scharr was among several WASP pilots in Orlando, Florida, who tested the physical effects of the human body's transition from sea level to high altitude (38,000 feet) in a pressure chamber from which oxygen was gradually sucked out. Scharr developed an embolism in her knee during one of these tests, even though she had been wearing an oxygen mask.[77]

WASP "Dot" Swain Lewis, 44-W-5, was initially assigned to Columbus Army Air Field in Columbus, Mississippi, where she tested the twin-engine AT-10. Later, based at the Laredo Air Force Base in Texas, she flew B-26s, trailing targets about 400 feet behind her airplane so that gunners in airborne B-24s could practice their shooting.[78] Lewis comments in *How High She Flies* that she faced more danger from a potential collision with other airplanes also flying target practice than from being shot at. Lewis also tested P-63 and P-40 fighters.[79]

At the Las Vegas Army Air Field, Marie Mountain Clark taught instrument flying and worked on range estimation exercises for gunnery students tracking targets.[80] This required her to make shallow dives toward a building at 180 miles per hour, a duty some pilots considered boring or dangerous; but it was one that suited Clark because she considered herself a well-coordinated pilot with good judgment and liked the chance to practice precision flying.[81]

Byrd Granger shares a few somewhat obscure facts about the scope of WASP duties in *On Final Approach*, such as training on experimental gliders — which were then considered a feasible means of transporting troops behind enemy lines — night flying, delivering soon-to-be-junked "war weary" aircraft, and weather detail — one of the more unusual of WASP duties. WASPs transported military meteorologists who worked for the civilian Weather Wing. The women on weather detail stayed in hotels or apartments and they flew five or six days a week from bases located in the major weather regions of the continental United States. Granger writes that the women had to buy their own food, but, as civilians, they managed to avoid having to do the calisthenics associated with military service. WASP Babette DeMoe Edinger, 43-W-7, flew weather personnel from Santa Monica, California, to islands located far offshore, a place where navigation by radio communication was impossible.[82]

At the Las Vegas air base, Elizabeth Strohfus' wartime work was diving at B-17s in an AT-6 for (camera) target practice. During her time in the WASP, Strohfus flew as many hours as she could possibly get, and she often offered to substitute for pilots who couldn't make their flights at her air base. Eventually her enthusiasm for being airborne converged with her physical limitations, and she suffered an episode of flight fatigue which she writes was cured by two weeks of rest at her home in Faribault, Minnesota.[83]

In *Global Mission*, General Arnold recalls taking a working "vacation" in the High Sierras in 1944. He writes that a young WASP flew the courier airplane that successfully dropped a pouch of war intelligence intended for him into a clump of woods in the mountains, another aviation task performed by the WASP.[84]

Although several women tested the device, a women's relief tube system, similar to the one used by males during long flights, was ultimately deemed impractical and was abandoned. After Helen Richey graduated from the WASP training program upon her return from England, one of her assigned tasks was to determine the practicality of females using the relief tube that had been designed for males. Richey experimented with the device while flying formation with Teresa James in a P-47 aircraft. Although she was ultimately successful in her attempts, she reported that the contraption was a challenge to use, despite the fact that she'd had "lots of practice" with it during her ATA days.[85] Helen Harrison also mentions using a relief tube, but not very successfully.[86] One widely adopted tactic among the women was simply limiting their intake of fluids before flying.

Hazards and Sacrifices

Whether they flew in places that never experienced enemy attack or risked their lives every day flying in combat, the women pilots of World War II pursued a dangerous occupation. The relentlessly pragmatic logic of wartime dictated that the British Air Transport Auxiliary and the United State Air Force use tactics whose primary criterion was whether they enhanced the country's air superiority. In England, ATA pilots delivered every airplane manufactured to wherever it was needed. In America, women ferried most of the airplanes moved domestically during the war, but they also tested-piloted, transported personnel, instructed, and flew targets for gunnery practice with live ammunition. They tested new or damaged aircraft and experimental high altitude equipment. They flew "unflyable" airplanes to prove — to male pilots — that it could be done. The major reason that American and British women flew military aircraft at all was that by doing so they freed the men to fly in combat.

Flying While Expendable

Women pilots in England and the Soviet Union were treated the same as their equivalents, the male pilots in their organizations. The Women's Section in ATA encountered the same challenges that all ATA personnel faced. They were assigned no special duties because of their sex, and they were expected to follow the same rules. With the exception of Class VI flying boats, women flew every aircraft type used by the RAF, and, during the later years of the war, they earned the same rate of pay. Female pilots in the Soviet Union enjoyed the benefits consistent with their elevated status in the Red Army. Pilots received better, and more abundant, supplies than other aviation personnel or the infantry, although privations were common throughout the Soviet armed forces during the war. Soviet women pilots faced no unusual hazards, but they were not shielded from risk simply because they were female.

93

The Americans, however, did not enjoy such equality. Granger claims that black (male) pilots and (white) female pilots in America were both considered expendable. They were ordered to fly questionable aircraft in poor weather conditions and she writes that pilots in both groups were routinely overlooked when desirable assignments became available. Instead they flew surplus airplanes and performed other less than glamorous tasks.[1] American women pilots were expected to fly in bad weather, while their male counterparts were not.[2]

The WASPs were "guinea pigs" for new technology. It was not uncommon for them to be used as "examples" for male pilots who were unwilling to fly airplanes that they considered dangerous. The women's working conditions varied, but the premise — if a woman can do it, how hard can it be?— was a constant in the lives of female pilots. All of the women knew stories of unrecognized and highly skilled pilots who ultimately astonished their male peers when their sex was finally revealed.

Jacqueline Cochran's prewar bomber flight satisfied the needs of the U.S. Air Force because it brought publicity to the Lend-Lease program, and at the same time it was designed to boost the interest of male pilots who were not exactly clamoring to fly airplanes across the Atlantic. Cochran's flight served her purpose too, giving proof that flying a complex bomber was not beyond the capability of a woman. After the war began, Nancy Love and Betty Gillies also prepared to fly a Lend Lease B-17 from Canada to England. During the summer of 1943 General William H. Tunner, of the Air Transport Command, gave his permission for the two WAFS to deliver a B-17 Flying Fortress via an extreme northern route from Maine to Labrador to Greenland, and after a stop in Iceland, to land in Scotland. The venture was a bid to show reluctant male pilots that the rigors of the wartime cross-oceanic route were no big deal, particularly if two *women* could do it. The women's flight was aborted when Tunner's announcement to Brigadier General Paul E. Burrows came to the attention of General Arnold, who was dining with Burrows in London. Arnold ordered the two women grounded at once.[3]

The U.S. Air Force used women pilots to counter resistance from male pilots who were afraid to fly a particular aircraft. Paul W. Tibbets, who would later pilot a B-29 on the atomic bombing run over Japan, taught two WASPs how to fly the bomber, an airplane considered unflyable — by men — when it was developed.[4] The idea behind teaching women to fly the B-29 was that no self-respecting male pilot could claim that the airplane was too dangerous or complex if a female could fly it. Employing this strategy had the purpose of shaming the men into cooperating.[5]

Adela Scharr was the first woman to fly a Bell P-39 Airacobra, a fighter (pursuit) airplane that required a different method of flying than others. Pilots who failed to employ alternative tactics when taking off in a P-39 frequently met disaster. The airplane became known as the "Bell Booby Trap" or the "Flying Coffin."[6] To prepare herself, Scharr practiced on a Curtiss AT-9 trainer, and when she had mastered that aircraft type, she thought she would be ready to fly the sleek tricycle-wheeled fighter. Scharr listened closely to the stories of pilots who had flown the plane, absorbing their warnings about its idiosyncrasies. Meanwhile, accidents involving P-39s kept occurring. Scharr notes that there was no ground school yet for that type, so her formal instruction on the airplane was scant. Concerned for her safety, several P-39 pilots tried to convince her not to fly the plane at all. Scharr persisted and when the men remarked that the engine's coolant lost effectiveness during their flight checks prior to takeoff, she was struck by an insight — most flight check tasks ought to be done prior to starting up the rear-positioned engine because it failed to cool effectively while on the ground. Scharr wrote out all of her preflight procedures and then she rearranged the tasks on paper trying to devise a plan for avoiding delays in the checking process. The scheme she developed allowed for no dawdling while the engine was running. Scharr was among the very few pilots who never had a problem with the P-39, but she was diligent in her preparations and she understood the ramifications of the fighter's unusual design. Her method for taking off worked with the fighter's characteristics, rather than against them. She describes her first solo P-39 flight: "That was all there was to it. One solo hop. No mistakes. Congratulations, me. I was still amazed at how fast I was going when I landed. It was breathtaking, just like my first landing as a student which my instructor made in a Wright J-5 Travelair. Somehow I knew that this, too, like the experience in the Travelair, would be the only time I would get this exhilaration in a pursuit airplane. From now on it would be old hat."[7] After Scharr's first P-39 flight, an air force major remarked, "You know ... when the boys get into a town and let a girl know that they fly pursuit, they're looked upon as little gods. Now you'll come along and everybody will know they aren't any better than the other pilots. Even a woman can fly 'em."[8]

Weather

Weather conditions were an ever-present, and highly uncertain, aspect of the pilots' experience. During the war years, forecasting was imperfect and undependable. The best information about the weather that a pilot might encounter was whatever could be gleaned from someone else who had just come from a particular destination. Changes in atmospheric conditions presented hazards just as formidable as those from the enemy.

In the United Kingdom, radio transmissions from ships at sea were not permitted during the Second World War. It was too dangerous to risk enemy interception of radioed messages, so weather conditions to the west of the island nation were never well known. Under these radio blackout conditions, the ATA's meteorology office did the best it could. At their ferry pools, pilots checked each morning for an update on the day's weather. Each pilot was ultimately responsible for deciding whether it was safe to fly. Canadian Vera Strodl's greatest fear while flying with the ATA in England was not enemy attack, but the ever changing British weather. She says her experience in the Women's Section ultimately convinced her of divine intervention, and she resolved to spread the gospel after surviving several close calls in the air.[9]

Weather forecasting was primitive and inexact in the United States as well, but meteorologists worked with the best information they had. Besides that, prewar pilots had little or no experience flying in parts of the country outside the region where they lived. WAFS Teresa James remembers flying cross-country from the eastern United States to the West Coast: "My worst trip was the first one to California. I had never flown over mountain passes. It never occurred to me that it was hazardous."[10]

After graduating with WASP class 44-W-4, Frances Roulstone's name was drawn to be the woman who would copilot Jacqueline Cochran's airplane back to Washington, D.C., after she visited Avenger Field in Texas. Flying through an electrical storm on the way, the two pilots were well aware of how risky their situation was, but their two passengers appeared more excited by the scenery below than afraid for their lives. Cochran's airplane was out of fuel, so she landed in a field in Versailles, Kentucky, and housed them all with a friend until the weather cleared and her plane was checked.[11]

Rules and Regulations

Sometimes regulations designed for the purpose of limiting risk or ensuring the safe delivery of an airplane worked against an individual pilot's safety. The ATA rule forbidding ferry pilots from "flying blind" turned into a dangerous limitation when the weather unexpectedly deteriorated, making the ground below invisible to the pilot who was allowed to fly only within sight of the ground. ATA pilots flew without radio equipment because their airplanes weren't so equipped when they left the factories. But radios were not considered necessary for them in any case because they were forbidden to fly in conditions in which radio contact would be needed. Pilots were not to fly with less than 2,000 yards visibility and 800 feet of cloud base.[12]

Instrument flying instruction was deliberately omitted from ATA's training program, not only because pilots were expected not to fly in the conditions that necessitated blind flying, but also because they flew short range aircraft with a limited fuel capacity; so there would probably not be enough time to continue on to another destination, anyway. The ATA's rationale was that being ignorant of instrument flying would prevent pilots from taking undue risks and "going over the top" of the clouds.[13] In *Spreading My Wings: One of Britain's Top Women Pilots Tells Her Remarkable Story from Prewar Flying to Breaking the Sound Barrier*, Walker describes an impromptu lesson provided to her by two RAF pilots who taught her the basics of "blind flying" by explaining how aircraft instruments worked using diagrams drawn on a night club's pink tablecloth.[14] "If in doubt, bail out" were the words that stuck in Walker's mind the very next day as she flew over Little Rissington in thick cloud. Walker had hopped into her plane so quickly she had not had time to change out of her uniform skirt; therefore, she didn't want to parachute, at least not immediately. She flew as low as she dared and finally glimpsed a small grass airfield through the clouds. The field was soggy with rain and snowmelt, but Walker landed her Spitfire intact, the muddy water splashing into her open cockpit as she rolled to a stop. Walker recalls that an RAF Link (instrument) instructor who had seen her land and met her airplane on the ground was full of praise for her skill on instruments![15]

Insufficient Training

The training program conducted by the ATA has been credited by many of its members for the program's success and low accident rate. ATA's training program was typically considered excellent, but its pilots would no doubt have benefitted from learning to fly with instruments, just in case bad weather occurred. ATA pilots were forbidden to fly at night and they were expected to employ good judgment about flying if weather conditions looked like they might deteriorate. Walker, the recipient of the just-in-time lesson on flying with instruments, might have lost her life for lack of this particular aspect of flight training.[16] She comments: "ATA decided that their recruits, such as I, would not be taught to fly on instruments — for as we must stay in sight of the ground, we should be alright."[17]

Across the Atlantic, flight training provided for the WAFS and for the WASP was not equivalent. The differences between Love's and Cochran's plans for employing women as pilots guaranteed that the content of their instruction programs would be different. WAFS did not receive the kind of intensive training that the less experienced WFTD and WASP did. The one-month-long WAFS training course was based on the premise that highly experienced pilots required nothing beyond basic instruction in military protocol and a short course in flying "the Army way." But the training program's brevity ultimately impacted the women's safety. As Jacqueline Cochran noted in her *Final Report on Women Pilot Program*, the failure of the United States Air Force to administer a consistent training program for all women pilots resulted in a higher rate of accidents and deaths for the original WAFS than for the WASP.[18]

Cochran required that the women who had served in England with the ATA complete WASP training before they could fly in America. That rule was absolute, even for experienced pilots like Helen Richey, who already had twelve years of aviation experience and had flown for a commercial airline in the United States before flying in England. Cochran knew that flying military aircraft was very different from flying either private or commercial airplanes.[19] Richey may have chaffed at the requirement, but Cochran's statistical summary in her final report on the WASP program proves the importance of a rigorous and consistently applied training program for the pilots.[20]

Wartime Conditions

There were extraordinary precautions taken against enemy attack in England and America, and those precautions posed hazards to pilots too. In Great Britain, cities and aircraft manufacturing centers were protected by large retractable balloons arranged in barrages that could be raised or lowered, and which were repositioned frequently. Safe air corridors were maintained for friendly aircraft and information about them was available for ferry pilots to use when approaching the airfields. When the configuration of the balloons was changed, pilots were duly informed, but they were not allowed to carry maps or diagrams of the safe routes for security reasons. Veronica Volkersz comments on the balloons:

> Most disliked was the Liverpool barrage, through which we had to fly practically every day. Nearly all our work took us up North. Liverpool was always submerged in smog and we rarely saw the balloons, though we knew they were there. There was a corridor, not reassuringly wide, through which we flew. The drill was to fly to Sealand aerodrome, follow the road up to the Mersey, turn on to a course of about 050 crossing Speke until one saw the railway line, then turn due north and hope for the best. If we hadn't hit anything within five minutes, we were O.K.[21]

In densely populated Great Britain, unlike the wide open spaces of America or the vast expanses of the Soviet Union, roads and railroad tracks were threaded throughout. ATA pilots, who had learned to fly somewhere else, had to work hard learning to navigate from place to place in England, often by memorizing maps, which they had to carry with them but could not mark.[22] Losing a map was a breach of security. Pilots' personal codes were printed on them, so if a map was lost and recovered, its owner could be traced.[23] Canadian Marion Orr cites what she discovered were the two greatest hazards of flying with the ATA — the high likelihood of getting lost because roads ran everywhere, and bad weather, because the pilots received no instrument training.[24]

After the United States declared war on Japan in December 1941, American airfields and strategic factories were hurriedly camouflaged. The work of the Bureau of Commerce's air marking program, which had systematically identified airfields and towns across the country, was rapidly undone. Identification from the air could provide assistance to the enemy, but its absence caused American ferry pilots difficulty finding their destinations while airborne.

Mechanical Failure, Human Error, Sabotage

Mechanical failure played a part in the women pilots' accidents and close calls. Although the exact number of occurrences is unknown, pilots who experienced problems with an airplane occasionally attributed them to sabotage. Sometimes it was difficult to determine whether accidents were the result of slipshod maintenance or intentional damage. If sabotage was the cause, had it been perpetrated by the enemy, or by airbase personnel resentful of newcomers who happened to be female? American accounts cite suspected instances of deliberate damage, but no one discusses sabotage in memoirs about the ATA, nor are there direct references to it in the Soviet women's accounts.

Human error was always possible. A few women inadvertently joined the Caterpillar Club, its name derived from the silk fabric used in early parachutes. On August 23, 1943, during her WASP training, Marie Mountain Clark "joined" the club by taking an emergency parachute jump after being thrown from an open cockpit during a spin. Clark writes that she was embarrassed by her poor spin recovery, but she took great pride in her performance during the jump. In *Dear Mother and Daddy*, she includes a January 1944 letter from the Irving Air Chute Company, welcoming her into the Caterpillar Club, promising her a membership card and engraved pin, and asking her to send whatever information (clippings, etc.) she was able to provide about the incident. Clark never received the membership card or pin, attributing the company's omission to the confusion of wartime, the war likely greatly increasing the club's membership, but she kept the parachute's ripcord as a souvenir.[25]

WAFS Florene Miller was practicing "touch and go" landings in a P-47 as part of transition training late one November day, when the airport announced it was closing immediately because of hazy conditions. Miller was using the shorter of two runways at Love Field in Dallas because the longer one was full of airplanes, and the P-47, which landed blind because the pilot's view was blocked by its large engine directly in front, might hit them. Telephone lines were strung along the airfield's boundaries and she needed to clear them while staying low enough to use the whole runway and stop in time. In low light and thick haze, lower than she planned to be, she hit a pole on her way down. The P-47 nosed up and spun to the right. Miller, who had learned aerobatics in her early flying days, kept control of the shuddering airplane, circled around, and took a course north of the populous city, trying to decide what to do. It was November 1943, with darkness not far off. Her first thought was to bail out, but then she started musing whether she could save the airplane instead. She turned it around and headed back. An illuminated rotating red horse, an early Mobil logo, provided the only clue to her location. She searched for Love Field but saw only darkness. After many tries and no contact, her radio attempts were at last heard by a nearby mechanic, who replied to her calls and relayed her messages to Love Field by telephone. Miller had to know if the airplane still had two wheels, so she asked to make a pass over the airfield's tower to confirm. They thought so, was the reply. Then she asked for some light along the other runway, the longer one, full of airplanes. Two jeeps were positioned to illuminate Love Field and policemen stopped nearby traffic so as not to interfere with their headlights. Miller made one pass to orient her approach and then she landed. She was frightened, but worried too that she'd disgrace herself by falling off the wing. She reports briefly considering running away because she was sure her commanding officer would terminate her, but he ran up with congratulations instead. The accident had ripped out the belly of the plane, damaged some of its instruments and a propeller, but left the tail wheel intact, giving her the ability to taxi straight along the line of airplanes. Miller went on to pursuit school and the city fathers in Dallas decided to bury the offending telephone lines.[26]

Byrd Granger recovered the stability of her airplane as it went into a dangerous secondary spin, a situation initiated by a flight instructor Granger considered incompetent and who subsequently quit teaching after the incident.[27] On the other hand, Jean Hascall Cole, 44-W-2, describes accidents that she thinks could be considered controversial which had occurred during her WASP training and later while the women were stationed at their air bases. At Sweetwater's Avenger Field, two advanced training aircraft, AT-6s, collided head-on while approaching the airfield from opposite directions. Two WASP trainees, Mary Hawson and Elizabeth Erickson, died in that accident.[28] Frances Roulstone, 44-W-4, tells of flying a PT-19 Fairchild trainer on her first solo cross-country flight over the Uvalde Mountains in Texas. There was no radio communication in the mechanically unsound airplane. Soon the weather turned bad. After leaving Sweetwater, Roulstone flew into a weather front, and her routine check of the airplane's instruments showed her oil pressure gauge reading zero. Panicked, her thoughts turned to the phrase "underneath are the everlasting arms," which she repeated over and over until she could finally see the airport at Brownsville in the distance. Signaling trouble by rocking her wings, Roulstone landed the plane just as its engine quit. The trainee avoided a court-martial over that incident only when an examination revealed that the airplane's engine log showed that it was unfit to fly. Indeed, Roulstone was commended for bringing the airplane in to the airport safely.[29]

Cole notes that it was not unusual for one version of an event to be widely accepted by the women themselves, while a very different version appeared in official documents. In her book, *Women Pilots of World War II*, three former WASPs, Lorraine Zillner, Mary Ellen Keil, and Leona Golbinec, describe occasions when each of them was convinced that her airplane had been damaged deliberately. Zillner parachute-jumped from an airplane that had its rudder cable cut; Keil successfully landed an airplane with a dislodged flight control unit, which had come loose as soon as she touched it; and Golbinec survived a landing performed with no aircraft controls at all, except for the throttle.[30]

Fagan cites what she refers to as "known instances" of sabotage in her book, *Zoot Suits and Parachutes*:

> All through training we had heard stories about sabotage on WASP planes, but when it happens to someone else, you discount the validity thinking it could never happen to you. Yet, there were a number of documented cases of sugar in the fuel tank; grass on one occasion. Mechanics in particular knew how to cross fuel lines with coolant and hydraulic lines — and did. Several accidents happened from just such acts and some were fatal. Tires were slashed just enough to get off the ground, but blow on a landing. A small vial of acid was found in a packed parachute. We were encouraged to learn to pack parachutes and could pack our own, but who had time for such things? We just had to be cautious and realize some of the weird things could happen to us. I never heard of anyone being nailed for sabotage. Time was too precious to waste on such small matters and WASPs were expendable.[31]

Fagan recounts a "practical joke" played on Ann Cawley O'Connor, 44-W-7, at the Stockton, California AFB. O'Connor noticed a red flag in the cockpit, indicating a "vital part missing," as she prepared a surplus AT-17 for transport to Blythe, California. She checked the airplane again and again, but found nothing amiss. Only when landing at a fuel stop in Victorville, at the end of an uneventful flight, did she recall the smug smile on her mechanic's face, watching her look for whatever might be wrong with the airplane.[32]

Camp Davis in North Carolina had already earned a reputation for inept and careless practices by the time Jacqueline Cochran chose the top pilots from WASP classes 43-W-3 and 43-W-4 to work there, towing targets for artillery practice.[33] Stationing WASPs there was Cochran's first attempt to expand the women pilots' responsibilities beyond ferrying. There were two crashes at Camp Davis, one month apart, which resulted in the deaths of two WASPs, Mabel Rawlinson and Betty Taylor Wood. Cochran visited the contingent of WASPs to learn from the women themselves exactly what had happened.[34] The women told her that they were flying airplanes that were combat rejects, passed on to the WASPs without the necessary repairs. The director of women pilots conducted her own investigation, but she preferred to keep the matter quiet, fearing that bad publicity might adversely affect the progress of her WASP program. Betty Greene and Ann Baumgartner Carl, both from class 43-W-5, were chosen to replace three WASP pilots who were no longer working at Camp Davis, including the two accident victims. Carl cites two major problems at the base. The first was that the WASPs had not been accepted, especially by the base commander, and the second that airplane maintenance there was poor, even nonexistent. In the case of "minor" problems, the custom had been to note the problem, but allow the airplane to fly. The cause of one WASP crash was traced to sugar in the airplane's fuel, while the second death was the result of a stuck canopy latch, preventing her from escaping the burning airplane on the runway.[35] Greene, in *Flying High*, mentions that Cochran's investigation uncovered evidence of sabotage — sugar in one of the airplanes' fuel tanks.[36] After those incidents, two of the WASPs working at Camp Davis resigned. Carl describes her experience:

> Duty at Camp Davis had few carefree interludes. Indeed, in many ways it was grim. But it showed the cool courage and dedication of the WASPs in their service in the Air Force that they faced the difficulties and dangers here without help from Commanding Officer Stephenson or Cochran or Washington, and took upon themselves the task of protecting themselves as best they could. Without complaints, they continued to fly missions for artillery men who were only just learning how to shoot their guns.[37]

6—*Hazards and Sacrifices* 103

Camp Davis ceased to be a WASP air base in March 1944. The women who'd been stationed there were sent to Otis Field in Massachusetts and to Liberty Field in Georgia.[38]

Granger describes an instance of sabotage in *On Final Approach*, in an entry dated December 1, 1943: "Then there are those tales coming out of Wichita where weathered-in WASPs are given a tour of the Cessna UC-78 production line. One watches a heavy-set woman, a gum chewer, pop a rivet, look around to be sure no one sees what she does, then takes gum from her mouth and smoothes it in the hole left by the rivet."[39] Granger elaborates on its scope in her next paragraph:

> Fatalities among ferry pilots are high. Some are attributable to careless production. Far more go down sabotaged. Every plane lost, every pilot killed is a plus for the Axis. A disgruntled employee hired to safeguard planes on a factory field pulls up grass by the handful and stuffs it into a fuel tank. Traitors in the ranks of mechanics cross fuel lines with coolant lines. Sugar is poured into fuel. Tires are expertly slashed, not to blow on takeoff, but on landing. Special Safety Bulletins appear so often that pilots tend to ignore them. One

After two fatal accidents involving WASPs stationed there, Jacqueline Cochran visited North Carolina's Camp Davis to hear about the women's working conditions firsthand (The Woman's Collection, Texas Woman's University).

advises that parachutes be inspected very frequently. A WASP carries hers to the Parachute Department and watches as it is laid out on a long table. Sure enough, tucked into it is a loose-stoppered small vial of acid. Holes in a chute don't improve a free fall from an airplane. Week by week tension mounts.[40]

Flying in Combat

Soviet women pilots had three different roles in the Red Army. They were fighters, day bombers, and night bombers. In combat, fighters flew alone, but usually in formation with other airplanes whose pilots served as wingmen in combat situations. Bombers flew with a crew aboard, including a navigator, but the solo flyers of fighter aircraft needed skills in piloting and navigation. Pilots flew day and night in all seasons and every kind of weather.

When their training period was over, the three women's regiments moved to their assigned locations; the 588th Night Bomber Aviation Regiment went first to the Southern Front (Stalingrad), as did Raskova's 587th Day Bomber Aviation Regiment. None of the regiments remained in their original locations, however. By the time the Great Patriotic War ended, the 588th had traveled as far as Germany and the 587th was located in Lithuania. The 586th Fighter Aviation Regiment, which defended fixed targets such as cities, airfields and transportation hubs, went to Saratov, southeast of Moscow, and over the course of the war the regiment had moved as far west as Vienna.[41]

At these locations, the women's living conditions varied depending on their proximity to active combat and to the time of year. Accommodations for Red Air Force pilots were rudimentary but they were better then those provided to other recruits. The women often slept outside in summer, so as to be ready to fly at a moment's notice; but in winter, they bunked in earthen dugouts heated by small stoves. The regiments relocated whenever shifting battle lines made moving necessary. Yevgeniya Zapol'nova, an armament mechanic in the 125th Guards Aviation Regiment (originally the 587th), describes their daily life in field conditions. Summertime allowed the women to sleep under their aircraft wings, but the arrival of winter necessitated that they stay in dugouts that were subject to flooding or becoming snowbound. When fuel for their stoves was in short supply, it was hard for them to dry out damp clothes and boots.[42] Women in the 586th Fighter Aviation Regiment experienced similar living conditions when they were stationed in Anisovka. The rustic encampments ensured that the women suffered through rodent infestations that were particularly severe in the fall and winter of 1942 in the Volga region. Evgeniia "Zhenia" Prokhorova, of the 586th regiment, while returning from a nighttime training flight, seemed to be having trouble with her airplane; its frequent side-to-side rocking, as it approached the airfield, mystified observers. Just after landing, Prokhorova dashed from the cockpit as if her airplane were only moments from exploding. It wasn't mechanical malfunction. During her flight several mice, cozily sheltered in the cockpit, tumbled onto her face and neck. Although Prokhorova was an ace pilot who was fearless in combat, rodents terrified her. The arrival of summer brought different conditions. It was then that swarms of mosquitoes plagued the women. While they lived in the field, a bath was a rare luxury.[43]

Women pilots who flew in the Soviet Union's Great Patriotic War occupy a unique category all their own. The number of women who could fly airplanes in the U.S.S.R. was larger than in any other country at the war's beginning because Stalin's government, which bestowed constitutional equality on females, encouraged women to join flying clubs. The pool of experienced female pilots exceeded the numbers that were generated by either England or America. Pennington notes that in 1941 there were between 100 and 150 air clubs in the Soviet Union and from one-quarter to one-third of the newly trained pilots were women.[44]

Women were not conscripted as men were. They served as volunteers in the armed forces, but it still took concerted effort by a famous aviator to bring them the opportunity to fly military aircraft. All of the women in Raskova's regiments flew combat missions. One regiment, the 46th Guards Night Bomber Aviation Regiment (originally the 588th), remained exclusively female from its formation until the end of the war. A few women pilots were already serving in other Red Air Force regiments when Raskova formed her aviation group, and many chose to stay in their original units rather than join the women's regiments. Some women who had begun their service in men's regiments were later transferred, at Raskova's request, into the all-female regiments.[45]

The Fighter Pilots (586th)

The 586th Fighter Aviation Regiment earned neither special appellation for itself nor Hero of the Soviet Union award for any of its members. Yet they performed admirably, making over 9000 flights and engaging in 4419 combat missions, which brought down thirty-eight enemy aircraft.[46] The explanation for why these fighter pilots won little recognition has been attributed to an omission by their first commander, Tamara Kazarinova, who was reportedly a stern taskmaster and never personally flew the modern fighter aircraft her regiment used. For that reason, she apparently failed to inspire respect from the pilots under her command, and discontent ran high among the women she commanded. Kazarinova's reassignments of personnel seemed to target only the disgruntled members of her regiment while showing little evidence of military logic.

The 586th was augmented by a squadron of male pilots who filled the ranks after those reassignments. Major Aleksandr Gridnev became the regiment's second commander after Kazarinova's six-month tenure. Gridnev sensed hostility from Kazarinova throughout his command, and he presumed that she didn't want the regiment to be better under him than it had been when she was its leader. Although the women had gotten notification of their award and they had already been issued special Guards clothing and could rightfully expect to receive official notice of Guards designation, Gridnev claims that Kazarinova was responsible for the fact that they were never formally awarded the status, because Kazarinova had not transported the required documents to Moscow.[47]

Pennington describes how Klavdiya Blinova's airplane was shot down on a mission with the 586th, but she managed to parachute into a ploughed field and was then captured by the Germans. Blinova was loaded onto a train with other captured Soviets. After two days she and her fellow captives escaped by using a penknife to raise a door bolt in their boxcar. The group wandered about for two weeks in the Bryansk Forest, located along the Desna River in the western part of Russia, before they found and rejoined their units.[48]

The 586th regiment was responsible for patrolling and defending fixed targets, protecting troops and escorting bombers, ground attack aircraft and personnel. Somewhat fewer than half its flights were combat sorties. Gridnev remembers that the women in his regiment received no special consideration because of their sex. In his recollection, he says that he would have liked to show them a more "tender attitude."[49]

The Dive (Day) Bombers (587th)

Marina Raskova assumed command of the 587th Day Bomber Aviation Regiment, but she was killed flying in a heavy snowstorm while trying to join her regiment at the front. Raskova never lived to see combat. Valentin Markov replaced Raskova as commander of the 587th in 1943 after she was killed.[50] Markov remembers his own reaction when he learned he was being reassigned to a women's regiment: "As I strode down the long corridor, I kept encountering old friends. When asked where I came from and where I was going—I would throw up my hands: 'Better don't ask. I am off to a women's regiment.' Friends pitied me openly."[51]

The women of the 587th, who were still mourning their beloved commander, were equally reluctant to accept Markov. When he arrived, they were away on a mission, but when they returned, their new commander noticed that they landed competently. However, proceeding to manage them in his usual manner, he brought tears to the eyes of a female armorer when he chided her for using excessive grease on the machine guns. Her reaction unnerved him, Markov recalls. Worried that his new regiment was inadequately trained and also that they might not accept his orders in combat, Markov required the women to practice formation flying, dive bombing, and high altitude flying; and he established the tradition of landing in a low level air-show formation after every victorious sortie to show that all had successfully returned from the battle. The women had trained on the Su-2 aircraft, but the regiment switched to the new Pe-2 bomber before going into combat.[52]

In time, their new commander came to acknowledge that the women who had volunteered for Raskova's air regiments were performing valiant service. He realized further that, for them, the very worst punishment possible was being grounded. Sometimes a men's regiment shared their airfield. Markov recalls hearing one of their commanders express embarrassment that his own pilots did not show the high level of skill demonstrated by the women. Markov's superiors made no distinction whatever between men's and women's regiments, placing both sexes wherever they were most needed, and the women took great pride in that fact. As their commander, Markov regretted sending them into such hellish conditions. In his reminiscence, Markov mentions by name the nine women who were killed while serving in his regiment during the war. His change of heart about the women he commanded is obvious in the title of his own essay in Cottam's *Women in Air War: The Eastern Front of World War II*: "I am Proud of my Regiment!"[53]

The Night Bombers (588th)

Unlike the ATA or the WASPs, who folded their wings at sundown, the women known to the Germans as *nachthexen* (night witches), flew throughout the hours of darkness, seizing the opportunity, especially during long winter nights, to disturb the sleep of encamped Germans. The 588th Night Bomber Aviation Regiment flew more than 24,000 combat missions, according to Pennington.[54] Interviewed for *Remembering War: A U.S.-Soviet Dialogue,* which commemorates the fortieth anniversary of the end of World War II, Yevghenia Andreyevna Zhigulenko recalls flying missions with her night bomber comrades for 1100 nights.[55]

The women of the 588th Regiment flew aircraft that Zhigulenko terms little more than "flying kites," small open-cockpit biplanes, U-2s, which were later known as Po-2s, constructed of canvas and plywood and which were so slow they could fly in relative safety only at night. Under cover of darkness, the airplanes were somewhat protected from German percussive rounds, which would have been enough to bring them down. If they were discovered by searchlights on the ground, the Po-2s provided airborne enemy fighters an easy target. In *Remembering War*, Zhigulenko describes the night her airplane and other bombers in her formation were caught in German searchlights. She evaded their beams by diving to the right, but she circled around to drop her bombs before returning home. Eight other airplanes and their crews, who were unable to escape, were lost.[56]

Larissa Litvinova Rozanova, flying a Po-2 bomber with navigator Nadya Studilina, recounts the experience of being caught in German searchlights and accosted by enemy fighters on a bombing run over the Cossack village of Krymskaya. That moment Rozanova told herself, "You can't very well turn back with your bombs just because you are afraid.... Well then, keep on thinking."[57] She watched as her friends in their own airplanes were targeted by the lights, attacked by fighters, and brought to the ground in flames. She decided to approach her target at treetop level, below the recommended altitude for the instantaneous bombs that she carried. To avoid detection, she turned off her engine and glided silently until after she had dropped the bombs. The shock wave from their impact at close range pushed her plane violently upward. She applied the throttle, an action that immediately drew machine gun fire from the ground and an attack by an airborne German fighter who missed when he shot into her cockpit. Every other airplane in Rozanova's original formation was lost that night. Only her tiny plane made it back, damaged by the bomb repercussion and pierced by bullets. She sums up the experience: "We've managed to outwit the enemy and accomplished our mission, but the loss of our dear girlfriends, the full measure of horror we experienced continued to haunt us for a long time. Yet subsequently we did square our accounts with the enemy, for the deaths of our comrades, our ravaged country, and the disruption of our personal lives."[58]

The women flew tiny airplanes with bombs attached to each wing and sometimes even wrapped around the bodies of the slimmest navigators. Some of the women would not use parachutes so they could carry even more bombs, but that practice was not standard procedure, although both Tat'yana Makarova, a pilot, and Vera Belik, her navigator, flew without parachutes to take an extra load of bombs.[59]

Weather vied with the dangers of combat in posing hazards to the women. Galina Bespalova, a flight navigator in the 588th, describes how she and her pilot got lost after making a night bombing run in a snowstorm. Navigating by compass in poor visibility, the two women landed their airplane just short of a telegraph pole in a recently deserted German airfield that hadn't yet been occupied by Soviet forces. The weather cleared the following morning, giving them a chance to orient themselves and return to their airfield in time to carry out the next night's mission.[60]

The Role of Raskova

The Soviets entered the war with only the aircraft that remained intact after much of the Red Air Force had been decimated by the Germans in June 1941. Those losses notwithstanding, Marina Raskova tried to obtain the latest aircraft models in use for the women in her regiments, including some American aircraft types that had been sent to the Soviets after the United States extended its Lend Lease program to the Soviet Union. One former WASP who participated in delivering them remembers flying fighter aircraft that were destined for the Red Air Force on a route over the American territory of Alaska. The Soviets were meeting pilots flying American P-39s and P-63s in Fairbanks by the time the WASP program started and one woman tells about meeting Russian women pilots who were among those receiving the airplanes in Alaska.[61]

Raskova's lobbying efforts on behalf of her flight group required that she frequently return to Moscow to arrange for the delivery of aircraft to her flight group and one source notes that she had to fight for her regiments' right to be supplied with up-to-date aircraft. She also ensured that radio transmitters were installed in the Yak-1 fighters that would be flown by the 586th Fighter Aviation Regiment.[62] She also prepared the course of flight training that was suitably rigorous for the combat conditions in which they would soon find themselves. The film *WASPs and Witches* depicts her seated at a desk and making speeches on behalf of the women.[63] A talented musician who had once contemplated becoming a professional performer, Raskova seized every opportunity available to sing or play piano while she lived at Engels. At Engels, Raskova trained on the Pe-2 dive bomber in preparation for her role as leader of the 587th Day Bomber Aviation Regiment.[64]

Raskova's patience and kind manner built loyalty in her volunteers, as did her refusal to accept the standard privileges of her rank as a major in the Red Army. The women in her 587th Day Bomber Aviation Regiment were devastated to learn of her death in January 1943 while en route to the front in a heavy snowstorm. They were determined to validate their commander's trust in them and to earn the right to bear her name. The 587th regiment earned the honorary designation of Guards in 1943 and its name was changed to the 125th M.M. Raskova Borisov Guards Dive Bomber Regiment.[65]

Flying in Stalin's Soviet Union

The Stalin era engendered a culture of suspicion in the Russian people. His Reign of Terror impacted the Red Army, including the women who had volunteered for service in the air regiments. Morale for soldiers and officers alike was low, conscription (for males only) was forced, political purges were frequent, and paranoia high. Anti-Semitism and other ethnic hatreds were common, both within and outside the Soviet Union. The extreme deprivations suffered by the Soviet population in the interwar years persisted into its Great Patriotic War against Germany. The Red Army received faulty equipment and its soldiers lacked adequate supplies. The Red Air Force was chronically short of aircraft, particularly after the surprise German attack had destroyed so many airplanes on the ground.

The aviation regiments engaged the Germans in battle, but at the same time they coped with a harsh political context in which Soviet soldiers might be executed by their own government if they were found to be captives of the enemy. The act of surrender was considered traitorous and those who were captured by the Germans were automatically accused of treason. One source alludes to the punishment that was imposed for losing a crew in battle.[66] During the course of the war it was common for the women pilots to return to their units and continue flying even after they had been injured, an action that can be attributed as much to their country's penalties for desertion as to their desire to fight their country's invaders. Besides their fear of being accused of treason, Soviets feared capture by Germans because of the deep hostility that existed between them, feelings based on long-standing ethnic hatreds. The Germans reserved the most severe treatment for their Russian captives.[67]

The story of Anna Timofeyeva Yegorova illustrates. Yegorova was the only female among the pilots in her regiment, the 805th Attack Aviation Regiment, 230th Division.[68] In August 1944 she led a formation of aircraft on a counterattack mission against a German position east of Warsaw at the Magnuszew Bridgehead on the Vistula River.[69] Her IL-2 Shturmovik, a fighter airplane often called a "flying tank," was hit and her gunner killed. The other pilots in the formation who returned to the air base reported her lost and a death notice was sent to her mother. Yegorova, who'd been badly burned after her airplane caught fire, was injured again parachuting from the burning plane at low altitude. She was captured by Germans who superficially treated her wounds and then imprisoned her for several months in the Kustrin POW camp, Stalag III-C Alt Drewitz for Allied soldiers.[70] Yegorova was liberated by her countrymen on January 31, 1945, but she was soon arrested by Soviet NKVD internal security troops on a charge of treason. Yegorova was interrogated repeatedly by government officials. She was denied her Party membership and stripped of her awards—a Party membership card, two Orders of the Red Banner, and a medal "For Bravery"—cherished items that she had managed to preserve throughout her months of captivity.[71]

The Soviet government's stance toward Russian prisoners of war was codified in Stalin's Order 227, issued July 28, 1942, at a time when his country's defeat seemed inevitable. The order, which all Soviet troops learned by heart, forbade retreat by the Red Army. Blocking units were deployed to the battlefield behind front line troops, with orders to execute anyone who disobeyed the command. An alternative to execution was to join the *shtrafniki*, penalty battalions whose members could atone for their "crimes" by carrying out missions so dangerous that they were essentially suicidal.[72] Stalin's "Not a Step Backwards" command implied that anyone in the Red Army found alive in a German prison camp was a traitor. The Soviet government claimed to have no prisoners of war, only traitors.[73] Yegorova writes that "spending even a day in German captivity stamped you with a mark that could never be washed clean."[74]

Awards for valor in battle were also subject to a test of patriotism. Yekaterina Zelenko's combat death is termed the only instance in which a woman pilot used her airplane to ram an enemy aircraft.[75] Pennington tells this pilot's story, adding a rationale for the Soviet government's reluctance to honor her bravery in battle. On September 12, 1941, Zelenko, a flight commander in the 135th Bomber Aviation Regiment, was attacked by two German Me-109s, leaving her wounded and killing her navigator. After ramming one German fighter, Zelenko's aircraft was shot down by another. Because her body was never recovered, the Soviet government refused to award her the HSU. Only in 1990, after eyewitness accounts verified that she had indeed perished in combat, was Zelenko named a Hero of the Soviet Union.[76]

Losses

Statistics from Curtis' *The Forgotten Pilots* show that 4.9 percent of ATA women pilots were killed during their service, compared with 8.1 percent of the men, who ferried airplanes for a longer period; 28.2 percent of the ATA women left the organization, compared with 29.9 percent of the men. Overall, "pilot wastage" (excluding American and RAF personnel) was 50.8 percent for women and 45.2 percent for men.[77]

However, ATA statistics that have been published in several other sources do not agree. Rosemary Rees du Cros, in *ATA Girl*, presents some additional statistics provided by Curtis: ATA's accident rate, as a percentage of deliveries made, was 0.39 percent during their worst winter (year not stated) and 0.16 percent during their best summer (year not stated). One-hundred twenty-nine men and fourteen women died while serving in the ATA, an overall casualty rate of 15 percent.[78] In *Harvest of Memories* Fahie cites a total of fifteen ATA women killed during the war.[79] And another source claims that 170 pilots of both sexes lost their lives in the ATA during World War II. The same source claims that the loss of life in the Women's Section ran approximately 1 in 10.[80] Mary Nicholson was the only American woman pilot who was killed serving in the ATA.[81]

Although numbers are hard to find, there are sources that state that approximately one-third of the Soviet women pilots were killed in the war.[82] Another source claims that, of the more than 1,000 Soviet women, who flew a total of more than 30,000 combat sorties, only a few more than *thirty* women were killed in combat during the war years.[83] However, this figure is hard to reconcile with the fact that it represents only 75 percent of the documented figure for American women killed during a considerably shorter period of service, and, while flying in noncombat conditions.

Yevghenia Andreyevna Zhigulenko, a pilot in the 588th Night Bomber Aviation Regiment, says that one-third of her regiment was lost during the war.[84] Figures collected by Pennington indicate that 29 women in the 46th Guards Regiment were killed, 28 died in the 125th Guards Regiment, and 10 lost their lives in the 586th Fighter Regiment. Pennington's figures count all regimental personnel — pilots and navigators, staff and ground crew, male and female.[85]

In the United States, thirty-eight women died in training or service with the WAFS or WASP. Ringenberg writes that eleven American women were killed during training.[86] According to Pateman in *Women Who Dared: American Female Test Pilots, Flight-test Engineers, and Astronauts, 1912–1996*, four WASPs were killed while testing aircraft in Texas, Arizona, and South Carolina. They were Betty Mae Scott, 44-W-3, Mary Hartson, 43-W-5, Peggy Martin, 44-W-4, and Jeanne Norbeck, 44-W-3.[87] Twenty-year-old Marie Michell Robinson, 44-W-2, died at the Victorville Army Air Field near the Mojave Desert in California. She was copiloting a B-25 Mitchell bomber.

Robinson was one of three women from her class who died while serving their country.[88] Most of the accounts of American women pilots during the Second World War list by name the thirty-eight women who died.

Sometimes, confirmation of a death could not be made until many years had passed. Gertrude "Tommy" Thompkins, 43-W-7, took off from Los Angeles, California, along with several other pilots on October 26, 1944. She was flying a P-51 Mustang in instrument weather conditions and she was destined for Newark, New Jersey, the embarkation point for fighter planes sent overseas. After Thompkins failed to arrive at her destination, a search yielded neither pilot nor plane. Although some sources continue to cite Thompkins' fate as unknown, one account tells the story of two hunters who discovered the wreckage of a vintage airplane in the mountains of California in 1985. "Tommy" was positively identified by the dog tags on her skeletal remains, still inside the P-51.[89]

7

War's End

The end of the fighting and the world's transition to peace brought inevitable changes for the women who had volunteered to serve their countries by flying military airplanes. The women's units were deactivated in an exactly reverse order from the sequence of their activation, beginning with the American WASP, which was disbanded in December 1944, and ending with the demobilization of the Women's Section of ATA, which was active until November 1945.

Leaders' Thoughts on Postwar Aviation

The leaders of the women's units held different opinions about how the war would affect future prospects for women pilots. Their projections ranged from sanguine to realistic. Nancy Love's postwar comments on a woman's future in aviation, presented in *New Wings for Women*, express considerable optimism. She claimed "There will be plenty of room for skilled women who have been trained in war time aviation and for many more new ones. Women pilots, having proved themselves, will no longer experience the opposition and prejudice of an earlier day."[1] Because Love considered flying an airplane to be quite easy, she commented that her peacetime ambition would be to encourage more women to learn to fly. She continues: "I firmly believe that, now the war is over, the small, easy-to-fly, privately owned ship will be commonplace and women will fly them as a matter of course."[2] In reality, Love was one of only a few women for whom such a smooth transition to peacetime aviation ever occurred. After she had fulfilled her commitment to WASP, she settled in Massachusetts, where she flew her own small airplane from her home on Martha's Vineyard to appointments on the mainland.

Jacqueline Cochran told her view of the future of WASP in a November 1943 speech at the *Herald-Tribune* annual forum on women in the aviation industry: "I can see them in the cockpit of many small commercial planes. I can see them at the controls of some of the small feeder lines. I can see them in the traffic control towers and they are ideal in the training of air students. But it is my considered judgment that when all is said and done, only about 25 percent of the WASPs will continue to earn their living in aviation."[3]

Marina Raskova died so early in her country's Great Patriotic War that she had scant opportunity to ponder postwar roles for women pilots. During her time as their leader, Raskova held the women in her flight group to high standards and she worked tirelessly; but, doubtless, she too looked forward to the end of hostilities when there would be time again to enjoy life in peace. When Raskova left Moscow with her aviation group, she placed her young daughter in the care of her mother. In speeches, her pride in the women under her command appears to fire her conviction that they would be glorified by future generations through epic and song.[4] Perhaps Raskova recalled the hero's welcome she and her two comrades had received after their momentous flight across the Soviet Union in the *Rodina*. That outcome never materialized for the Soviet women pilots.

The following pragmatic — and prophetic — words, attributed to Pauline Gower in Knapp's *New Wings for Women*, were written just after the Second World War ended: "I don't hold out any rosy prospect for women commercial pilots or engineers in this country now that the war is over. There are thousands of trained R.A.F. men wanting aviation jobs now that peace has come again to the world, and personally I doubt that women will have much of a chance in commercial aviation, as pilots or engineers."[5] Gower goes on to comment that women had not held these positions in aviation before the war began and they were not likely to hold them afterwards. She adds that this limited opportunity was no reflection on their ability as pilots but the result of economics and a new peacetime philosophy which would inevitably change their employment picture. Gower adds that women would still obtain jobs as air hostesses and find work in other nonpilot positions.[6]

Disbandment in the United States

The WASP as an organization ceased to exist on December 20, 1944, several months before the Allies declared victory in Europe in May 1945 and victory over Japan in August of that year. The campaign for militarizing the WASP began almost as soon as the organization was formed. One plan was to incorporate the women into the Women's Army Corps. Another, the plan preferred by Cochran and Arnold, was to place them into the Army Air Corps, in anticipation of its soon achieving independence from the army and becoming the United States Air Force.

Opinions about militarization were not unanimous, even among the women themselves. They disagreed about the prospect of becoming part of the army, and if that did happen, where they belonged in that branch of the military. The antagonism of Love's WAFS in the Air Transport Command/Ferry Division toward Cochran's WASP would eventually join forces with opposition from outside groups to augment the campaign against any kind of program for women pilots. Love herself opposed the WASP bid for militarization, as Granger notes in *On Final Approach*: "The situation is that Mrs. Love is showing up in Washington and goes from one office to another trying to sway the top brass against militarizing WASP."[7]

Georgia Congressman Robert Ramspeck, chair of the House Committee on the Civil Service, conducted an inquiry into the WASP in 1943, examining all of the program's expenses, from uniforms to officer training.[8] By the time Representative John M. Costello of California introduced a bill that would give the WASP military status in January 1944, HR 4219, it already faced serious challenges. The data collected by the Ramspeck Committee gave no support whatever. No longer encountering pilot shortages, the U.S. Army had phased out the Civilian Pilot Training Program in June 1944, and, with its losses in battle falling below anticipated figures, the army cut personnel. Male pilots returning home to the United States had few options for employment that would keep them out of the infantry and they resented women who held air force jobs that they wanted, particularly the pilots.[9] Many of CPTP's student pilots and instructors, now eligible for the draft, complained to their congressmen about the Costello Bill.[10] Columnist Drew Pearson fanned the flames of antagonism to the women pilots when he wrote that General Arnold, who supported WASP militarization, couldn't help being seduced by the ideas of the "winsome Jackie Cochran."[11] Editorial cartoons characterized the women pilots as nothing more than glamour girls.[12]

Of the many forces working against WASP militarization, not least was the fact that Jacqueline Cochran, the program's creator and tireless champion, was sometimes its biggest liability. Cochran objected to militarizing the WASP when it was formed because she didn't want her program to be part of the WAC and therefore under the command of Colonel Hobby. Cochran personally disliked Hobby; she never envisioned her pilots (or herself) under the command of someone from the *army* who admittedly knew nothing whatsoever about aviation.[13] Cochran preferred that the WASP enter the Army Air Corps, not only because she considered her young women pilots "a different breed of cat" from the WAC, but because she had no desire to lose her hard-won status as their commanding officer.[14] WASP Margaret Kerr Boylan, a contributor to Brinley's rewritten Cochran autobiography, provides a perspective on

Cochran's personality that was shared by many: "If she (Cochran) couldn't do it her way, she'd just as soon not do it anymore."[15]

A number of women later blamed the director of women pilots for the demise of the WASP.[16] However, Byrd Granger writes about being summoned to Cochran's office in June 1944 and asked to assume leadership of the WASP. Granger claims that Cochran was worried that the WASP militarization bill might fail as a direct result of negative feelings about her. According to Granger, Cochran was willing to sacrifice her command if it would help to pass the bill, saying, "What it comes down to, really, is that as long as I am Director of Women Airforce Service Pilots, WASPs will never be militarized."[17] Granger refused Cochran's offer.

In August 1944, Jacqueline Cochran had dinner with WASPs stationed at the Las Vegas AAB, where Marie Mountain Clark was located, and she chided the women for putting their personal lives ahead of their duty, reminding them why they had originally decided to participate. She supplied facts and figures about the WASPs supporting their outstanding safety record. She also told them the WASP bill was not "completely thrown out," but was sent back to committee. Cochran told the women that the alternative to the WASP "going in" would be to disband. She went on to describe the future role of the B-26 bomber and the possibility that WASPs might be trained to fly it.[18] Summing up her WASP experience, Elizabeth Strohfus, also 44-W-1, believes that Cochran did more for women in aviation than any other person.[19]

In *Sisters in the Sky*, Scharr writes that she had very little information about the campaign for WASP militarization at the time and explains why the women were not active in defense of their program, claiming they were simply too busy to attend to the news.[20] Clark reinforces Scharr's statement, saying that most of the information about the WASP bill came to her through her parents' letters because she and her fellow WASPs were too busy working to pay attention to the news. On June 22, 1944, however, Clark wrote to her parents from Officer Candidate School in Orlando, Florida: "You probably heard about our bill's being defeated by 20 votes in the house [U.S. House of Representatives] yesterday, but the opinion of the experts here is that it will eventually go through because General Arnold and the War department get what they go after. General Arnold is out of the country right now."[21] After the bill's defeat, Cochran proceeded with plans for WASPs to receive officer training in Orlando, Florida, during the summer of 1944. She wanted all the women she'd chosen for OCS to finish the course.[22]

The WASP training program continued to operate nearly up to the date of the program's demise. The film *We Were WASP* shows striking contrasts between the first and last graduation ceremonies. The first event in Houston depicts the women in crisp white shirts and khaki pants while Cochran wears a flowered dress under sunny Texas skies. The scene shifts to the "lost last class," 44-W-10, attired in neat blue WASP uniforms. At their indoor ceremony in Sweetwater on December 7, 1944, General Arnold pins wings onto each graduate's uniform while the film provides a background of upbeat big band music. The mood then changes. Somber-faced graduates of 44-W-10 sing WASP words to the tune, "Tell Me Why."[23]

General Henry H. "Hap" Arnold, an early opponent to WASP militarization, eventually changed his mind and testified before Congress in support of HR 4219.[24] Arnold became a champion of the program he had originally doubted. His speech to the last WASP class, graduating only days before disbandment, describes his change of heart:

> Frankly, I didn't know in 1941 whether a slip of a young girl could fight the controls of a B-17 in the heavy weather they would naturally encounter in operational flying. Those of us who had been flying for twenty or thirty years knew that flying an airplane was something you do not learn overnight....
>
> Well, now in 1944 more than two years since WASP first started flying with the Air Forces, we can come to only one conclusion — the entire operation has been a success. It is on the record that women can fly as well as men. In training, in safety, in operations, your showing is comparable to the over-all record of the AAF flying within the continental United States. That was what you were called upon to do — continental flying. If the need had developed for women to fly our aircraft overseas, I feel certain that the WASP would have preformed that job equally well.[25]

The documentary *Fly Girls* presents excerpts from Arnold's speech to the final WASP graduates in which he calls the program a "tremendous success."[26] In *Global Mission*, Arnold's account of his role as wartime commander of American air power throughout the world, he writes that "the WASPS [sic] did a magnificent job for the Army Air Forces in every way."[27]

Bea Falk Haydu, 44-W-7, who helped the WASP achieve militarization thirty years after their first attempt, notes that the women were not as aware as they should have been of the first campaign for militarization, or of the strong feelings against it at the time, because they were busy delivering airplanes. Haydu claims that WASPs only wanted to continue flying, pointing out that they had received identical training as the men, and after attending OCS "they came out officers."[28] Marie Mountain Clark claims that the seeds of WASP disbandment were planted when the organization was formed and placed into the Civil Service rather than the military, giving it a separate administrative category at the outset.[29]

A convergence of elements conspired against WASP militarization as the war was ending, but some scholars make the claim that the Costello Bill was defeated because female pilots threatened masculine "hegemony" more than enlisted women had ever done before, therefore the male-dominated powers (Congress, the press, the army) "squashed" them.[30] However, the story of the WASP bill should not be viewed so simplistically. Factors directly affecting the Costello Bill's outcome ranged from the unique origins of the WASP program itself to the timing of the women pilots' bid for militarization, which coincided with the return of men from service overseas in larger numbers than anticipated. Constraints that had been imposed by their own commander on WASP participation in the legislative process didn't help — Cochran had instructed the women not to comment. General Arnold's testimony on behalf of the WASPs was important, but many sources ignore his contribution. The defeat of the Costello Bill stemmed from a combination of two factors: the negative effect of Cochran's complex personality — she sought recognition, but always on her own terms — and the unfortunate timing of the bill's introduction to Congress.

Final Report on Women Pilot Program

Of the four women who made it possible for their contemporaries to fly military aircraft during the Second World War, only Jacqueline Cochran wrote about the experience, but her autobiography has to be read with more than a little skepticism because of its author's inclination to hone her carefully crafted public persona. In her 1954 account, *The Stars at Noon*, Cochran expended few words on the failed WASP militarization effort and instead described her desire to "do a little real fighting myself, either in a bomber or a fighter":[31]

> As the ex-head of the WASP, I also have the rank of Lieutenant Colonel in the Air Force Reserve. I would not have taken that job of head of the WAF if a full general's rank had been thrown in. If a fighting war should eventuate, I would, however, willingly lay aside my manifold civilian obligations, let my cosmetics business float again, and if necessary, in the lowest rank, crawl across the country on my hands and knees to be of aid to my country. I'll do this if I can't handle a fighter plane in combat. My inability to so fight in the last war was my great disappointment. A couple of times I thought I might get the opportunity to fly on a mission or two but nothing of this kind ever matured.[32]

Nancy Love, in her subservient role as leader of the women in the ferrying division, was not expected to report on the success of the women pilots' "experiment." It was Jacqueline Cochran, the director of women pilots, who wrote the *Final Report on Women Pilot Program* in the winter of 1944–1945.

Her first recommendation was that "any future women pilot program should be militarized from the beginning."[33]

In her report, Cochran notes that the experiment was designed to determine whether women were capable of ferrying airplanes or towing targets so that the men who had received combat training wouldn't be required to perform those tasks and instead could work in the areas for which they had been trained and for which only they were eligible. She comments that before the WASP was formed no one knew whether women could fly military aircraft. The only certainty was that a few women had qualified for licenses and some of them were outstanding pilots. Cochran adds that the WASP program was intended to develop aviators of day-to-day competence, rather than "hot" pilots. She describes the program's inception, recruiting and appointment, as well as its objectives, as follows: to see if women could serve as military pilots; to release male pilots for combat; and, to decrease the air forces' total demands on the cream of the manpower pool. Cochran elaborates on the requirements for entry, the curriculum, equipment, operation of training base, pay and living costs, organization and discipline at training base, formulation of overall WASP program, severances during training, severances during operations, hours flown, accidents, value of Army Air Forces training, medical features, a suggestion concerning any future program, militarization of WASP, inactivation of the WASP, list of WASPs in good standing, conclusions and recommendations.[34]

Cochran's report includes detailed statistics generated by the women's physical examinations during their application process and various numbers gathered throughout the program. She recommends that in future the women's age range should be adjusted downwards because the graduation rate declined and severances from the organization increased in direct correlation to a trainee's age. The report recommends that the upper age limit be reduced from age 35 to 27 or 28 years, and the lower age limit dropped from age 21 to age eighteen. Cochran cites the failure of the WASP to be militarized as the reason for its high rate of resignations. There are statistics proving that the highly experienced, but lesser trained, WAFS had an accident rate per flight hour that *exceeded* the accident rate of the less experienced, but more rigorously trained, WASP.[35]

Aftermath to Early Disbandment

Betty Greene writes that she was disappointed to hear the news of WASP disbandment, but she had already resigned the program in October 1944 when she had an opportunity to join the Christian Airmen's Missionary Fellowship.[36]

Margaret Ringenberg, Greene's classmate, was sorry that she had never advanced to ferrying pursuit planes during her WASP years, but time had run out for continuing the transition. Ringenberg found the cessation of the WASP program very depressing because she could find no opportunity to become a commercial pilot and was decidedly uninterested in working as a stewardess, the only career option readily available to women aviators immediately after the war.[37]

While she was still serving actively as a WASP, Marie Mountain Clark was posed the following question by her commanding officer — did she intend to keep flying after the war? Her answer would determine whether she trained on a P-39 if she chose recreational flying or on the much larger B-26 if she planned to fly professionally. Clark opted for the P-39. She writes that completing the United States Air Force flight training program had been one of most fulfilling experiences of her life and she felt deep sadness at the cessation of the program, because after that "most intensive, satisfying, and fulfilling" experience, disbandment was a great letdown.[38] Nevertheless Clark felt privileged to have had the flight training and she knew firsthand that her base in Las Vegas was a place where the WASPs were welcomed, respected, and, if they qualified, free to fly any aircraft on the field. The women stationed there were given time to fly and "check out" in as many aircraft as possible before they left.[39] Clark flew a P-39 one last time on December 20, 1944, feeling an acute sense of loss which she considered akin to bereavement.[40] Teresa James later compared her feelings about the disbandment to the emptiness she had felt while grieving her husband's death.[41] Adela Scharr writes that the return to civilian life after the WASP program ended was traumatic for some, and she cites the names of three women who had also flown with the ATA whose difficulty managing their postwar transitions ultimately resulted in suicide.[42] Ann Carl, the test pilot at Wright Field, comments: "So, with the WASP highly expert and often dangerous military flying still secret, in December 1944 we would turn in our equipment — uniforms, parachutes, jackets, oxygen masks, coats, boots, gloves — while we felt there was still flying work to be done. We would suddenly be civilians, without military benefits. Not even the G.I. Bill applied to us. The civilian instructors, however, were taken into the Air Force, individually, following the usual Ferry Command practice."[43]

Demobilization in the Soviet Union

The Great Patriotic War demanded every Soviet citizen's participation. At war's end in May 1945, Red Army regiments were demobilized gradually, giving time for all its widely scattered combatants to return home.[44] The Soviet government resumed its prewar stance regarding women in the military, and females were once again discouraged from pursuing military careers. Women who had volunteered to serve their country were expected to choose family obligations over a military career. The Red Army did not retain its women's units. Despite the Soviet Union's 1936 constitution, which proclaimed equality between the sexes, the unwritten policies of the Stalin era had far greater influence. During the war, Joseph Stalin had praised female partisans and war workers, but he did not acknowledge the women serving in the army or in Raskova's aviation regiments. Many women had been outstanding military pilots, but after the war Stalin proclaimed females unsuited to any kind of aviation, military or civilian. One woman pilot still remembers the hurt that those comments caused her.[45] Stalin also refused to allow women to participate in celebrations marking the war's end.[46] No official account about the Soviet women pilots' experience was written. In *Wings, Women and War*, Pennington cites one historian who has called them the war's "invisible combatants."[47]

Pennington writes, "The overwhelming message of Soviet propaganda was that although women could fulfill combat roles when duty called, they were not to expect permanent careers in the military. Their achievements were acknowledged during the war but then quickly forgotten or even obscured."[48] In memoirs, the women themselves state that they were war-weary by that time, and less concerned about limits on their activities than about rebuilding their devastated country. They had few resources left for fighting discrimination. Some suffered from health problems as a result of their wartime service. Those still recovering from battle wounds received medical discharges. It is still impossible to assess the performance of the Soviet women pilots, writes Pennington, because records about their activity are unavailable.[49] Evidence of their abilities as combat pilots remains anecdotal.

Some of the women from the aviation regiments managed to continue on active duty in the air force. Some returned to their civil aviation employment or worked in other aviation related fields. Others worked in science or engineering, or engaged in other professions, or they worked for the Party.[50] Before the collapse of the Soviet Union in 1991, women who had flown with the Red Air Force rarely spoke about their wartime experiences, and opportunities to interview them were few. However, when the Soviet archives were opened after political restrictions loosened, interviews with former pilots were easier to obtain and more information on their wartime activities became available. Members of the women's regiments felt more confident that they would not face retribution for talking about their experiences.[51]

Closure of ATA in Great Britain

ATA first officer E.C. Cheesman wrote an official account of that organization as it wound down in fall 1945. Knowing that the time for conducting interviews was growing short, Cheesman gathered memories from the participants while their recollections were still fresh, before they were discharged and scattered. Cheesman allows the ATA Women's Section due credit. The book's foreword by Lord Beaverbrook states that "the limelight of the war did not fall" upon the men and women of ATA, but a debt is owed them nevertheless. Cheesman devotes one chapter to each major ferry pool, including Hamble, the all-women pool. Another chapter tells the story of the Women's Section.[52]

The Hamble ferry pool closed in August 1945. In September of that year, the ATA staged an air display and pageant to benefit its benevolent fund.[53] Ferrying flights were becoming less frequent and the lighter schedule gave ATA pilots a chance to think about their futures. Rosemary Rees du Cros continued to ferry aircraft after the ATA officially ceased operations, working at the White Waltham ferry pool, the last one to close — in March 1946. Du Cros ferried two of the first jet aircraft, a Meteor and a Vampire.[54] Lettice Curtis, who had been in ATA since 1940, ferried military aircraft until the program's official end on November 30, 1945.[55] For his book, *Spitfire Women*, Whittell interviewed many of the women who had flown for the ATA and several stated frankly that they would have liked to have flown in combat.[56]

Jackie Moggridge, who had married during the war, was pregnant by November 1945. Nevertheless, she carried on with her work until ATA's closure precluded further flying. Moggridge says her physical condition helped to ease the disappointment she felt at the prospect of never again flying military aircraft. It wasn't until the day she departed White Waltham that she announced her impending motherhood to her ATA colleagues. In *Woman Pilot*, she writes that "My last memory of the A.T.A. is the look of startled incredulity on the faces of the men-folk as I drove away in a taxi."[57]

Pauline Gower died in January of 1947, shortly after giving birth to twin sons, leaving the task of compiling a history of ATA's Women's Section to Lettice Curtis. Curtis claims to have felt inadequate to the task, knowing that Gower should have been its rightful author; but she set to work compiling her own account of the creation, performance, and disbanding of the ATA organization in *The Forgotten Pilots*, which was published in 1971. Because her experience in ATA was as a member of the Women's Section, Curtis notes in her introduction that she has made the "theme of the book the women's side of the organization."[58] She assesses her personal reaction to ATA's closure:

In the event VE-Day did not feel particularly like the happy and glorious day to which we had aspired for the best part of six years. It proved in fact to be something of an anti-climax taking away incentive, our very *raison d'etre*, and putting nothing in its place. Moreover the unpalatable fact had now to be faced that within all too short a space of time, one would in order to exist have to find another job and in many cases another roof. To those of us who had nothing to go back to and nowhere particular to go (my mother had died during the war and the family home was now broken up) the end of the war was about as climacteric an experience as the outbreak.[59]

Commendations and Recognition

During the war, some of the American WAFS and WASPs were recognized for ferrying unusually long distances. Barbara Erickson, the WAFS commanding officer at Long Beach, California, once flew four 2000 mile flights in a five-day period. For that feat, she was awarded the Air Forces Medal.[60] Colonel Henry H. Arnold presented the Distinguished Service Medal to Jacqueline Cochran in August 1945.[61] Nancy Love was awarded the Air Medal for her "Operational leadership in the successful training and assignment of over 300 qualified women fliers in the flying of advanced military aircraft." Both Love and her husband, Robert, were honored on the same day when Robert Love earned the Distinguished Service Medal.[62] On July 1, 2009, President Barack Obama signed PL 111-40, the law awarding the Congressional Gold Medal to the WASP.[63]

In 1938, after their flight in the *Rodina*, navigator Marina Raskova, with pilots Valentina Grizodubova and Polina Osipenko were awarded the title Hero of the Soviet Union. It was the highest honor awarded to Soviet citizens, or to foreigners, and its accompanying gold star medal distinguished these heroes from any other award recipients.[64] Although most Soviet women awarded the HSU were airwomen or Resistance activists, Cottam writes that there were probably fewer awards made than women whose performance deserved recognition. She notes a reluctance to show favoritism among their commanders, especially if the woman's superior was a male who might worry that his own superiors would presume that a "special relationship" had existed between him and the recipient. Cottam thinks this is the reason that the commanders were not "overly generous" in awarding the HSU to female soldiers.[65] Pennington writes that 38 women aviators, of whom 25 were pilots from five separate regiments were awarded the HSU for their service during the Great Patriotic War.[66]

Two of Raskova's regiments earned the honorary title Guards for valor in combat. The title was awarded to elite units and traced its origins to tsarist tradition. Raskova's own regiment, the 587th Day Bomber Aviation Regiment, was determined to honor her after she died by earning the Guards distinction. In time they became the 125th M.M. Raskova Borisov Guards Bomber Regiment. The 588th Night Bomber Aviation Regiment earned the title 46th Guards Night Bomber Regiment in 1943 and then, because it performed so valiantly in battle over the Taman Peninsula, was renamed the Taman Guards Night Bomber Regiment. Sixteen women died in air battles there in the summer and fall of 1943.[67] Twenty-three women in the 46th Taman Guards Bomber Regiment were awarded the HSU.[68]

Only one of the three women's regiments retained its original appellation. Pennington notes that, despite having an unusual number of highly skilled pilots, the 586th Fighter Aviation Regiment never won an official Guards appellation, nor did any of its members receive an HSU award.[69] Some individuals were honored with other awards, such as the Silver Cross of Merit presented to former German prisoner of war Anna Yegorova in May of 1945.[70]

In Great Britain, Pauline Gower was awarded the M.B.E.—Member of the Order of the British Empire—during her service with the ATA. She was appointed a director of BOAC in May 1943, and her first task was surveying the British air routes. She was accepted into the Worshipful Company of Pattenmakers in London, the first woman ever so honored, in June 1943.[71] According to Fahie, Margot Gore, the commander of the second ferry pool for the Women's Section, was also awarded the M.B.E.[72]

Third Officer (Miss) M. Shiel is among the ATA pilots who won Certificates of Commendation, according to Appendix A in Cheesman's *Brief Glory*, for an act of outstanding merit during their service with ATA.[73]

Conclusion

Women who had served their countries as pilots during the Second World War faced rejection when they applied for jobs with the commercial airlines afterwards, despite their wartime aviation experience. Service in the military has been the traditional way to gain the training and experience needed to work as a commercial pilot, but the women's experience in military aviation did not lead to opportunities for them in the postwar years. Even highly motivated female pilots had difficulty pursuing a postwar career in aviation. Nevertheless, a woman who was determined to fly could usually manage to do so, but often that meant seeking creative paths to a career in aviation.

Postwar Aviation

United States

After World War II ended, there was little chance for American women to fly in the military. In 1948 the women who'd been WASPs were offered nonflying reserve commissions in the U.S. Air Force as second lieutenants. Some of them accepted, but others, finding these posts irrelevant after their years of active flying, did not.[1] Ann Carl writes that after disbandment, one hundred-fifty WASPs signed on for nonflying duty with the air force, two WASPs joined the army, two joined the navy, and one went to the RAF.[2] Jean Hascall Cole writes that postwar options for WASPs included joining the reserves in the U.S. Navy, the U.S. Army Air Force, the U.S. Marine Corps, or the Royal Air Force.[3]

A perusal of Turner's *Out of the Blue and Into History* proves that women who wanted to make a career in aviation after the war usually found a way, in spite of the near certainty of rejection by the commercial airlines. Turner gathered 673 accounts by or about former WAFS or WASPs, the most complete and current compilation of the women's postwar activities. Margaret

Ringenberg joined the United States Army Reserve and was commissioned a first lieutenant. However, she was discharged as soon as the army discovered that she had a dependent child. Ringenberg continued flying as an instructor and as an independent contractor. She enjoyed competition and participated in air races, in 1957 entering her first AWTAR, the All Woman Transcontinental Air Race, aka Powder Puff Derby.[4] Most of Ringenberg's postwar aviation activity involved competing in air races, including one around-the-world race in 1994.[5]

Elizabeth Strohfus comments that her WASP deactivation was the beginning of three years of "unsettled times," during which she married and lived a "normal life"—without flying—until the day she was invited to ride in an F-16 in August 1991. Strohfus even seized the opportunity to fly the jet herself—at age seventy-one![6]

Some women never flew again after their WASP years, which was as likely to be a voluntary decision as an involuntary one. Florene Miller Watson, the pilot who had managed a successful emergency landing on a darkened Love Field with a damaged airplane, pursued different interests as a wife and new mother after the war. She didn't fly after the WASP program ended and she notes that the experience of flying military airplanes was only one of her interests and she'd already done it. She knew that the demand for instructors was no longer great and decided that it was time to move on to other things. She chose to stay home, commenting, "I have always been a lover of the home." But other women purchased their own airplanes so they could continue flying in some capacity. Finding prospects for careers in aviation limited, some found jobs in related fields or as flight instructors, or they obtained immediate temporary work ferrying surplus World War II aircraft wherever those airplanes were needed, including delivering them to Mexico.[7]

Madeline Sullivan 44-W-2 comments on the difficulties of a female's obtaining a commercial pilot's position, quoting her boss's response to her desire to quit her job as copilot to an incompetent pilot: "Don't [quit], we've got so many complaints about this guy, he's going. There's nothing we'd like better than to give you the left seat [pilot], but I'm afraid it won't work."[8] Sullivan stayed in her position, noting that she did not "press the issue" because she felt lucky to have a copilot's job at all when there were still so many people who opposed the very idea of women flying. Later Sullivan travelled to Moscow where she met some of the Soviet "night witches."

Having no opportunity to fly commercially, yet reluctant to return to flight instructing on single-engine aircraft, Marie Mountain Clark decided that renting small airplanes to get flying time was not worth the money. She flew only briefly after the war, and once ferried war surplus planes from Oklahoma to Des

Moines, Iowa, in March 1945. Clark held both a commercial pilot rating and an instrument pilot certificate and had nearly 1000 hours of flying time in what she terms "aircraft of major significance" when she left the WASP.[9]

Gini Dulaney 44-W-2 whose husband had a postwar assignment in Hamburg, Germany, learned the techniques of gliding in a club made up of former Luftwaffe pilots.[10] Dulaney set a glider distance record while seeking out the thermals that provide lift to nonpowered aircraft, and in 1957 she beat Hanna Reitsch's altitude record by gaining 28,000 feet over her release point, as recorded on a sealed barograph.[11]

In *Zoot Suits and Parachutes*, Fagan writes that women who wished to pursue a career in aviation needed to become entrepreneurs and creatively made their own opportunities. Fagan did some instructing after the war and eventually she opened a GI Bill-approved flying school, the Lind Flying Service in Eastern Washington State. In 1948, after moving to the U.S. territory of Hawaii, she was turned down for a job as copilot for Trans Pacific Airlines because the vice-president of the company was convinced that no one would want to fly on an airplane with a woman in the cockpit! Fagan returned to instrument instructing for TPA and she taught at the Hawaii School of Aeronautics in her spare time.[12] Turner adds more about Fagan's postwar experience, noting that Jacqueline Cochran started an Aloha Chapter of the Ninety-Nines in Hawaii in 1950. Fagan continued to fly privately and she purchased airplanes of her own, one of which, the Piper Arrow, reminded her of a mini AT-6.[13]

Betty Greene continued to work as a pilot after her WASP service, leaving before disbandment in 1944. Greene considered her military experience to have been useful preparation for her work with the Mission Aviation Fellowship (MAF), an organization she helped create. While working as a missionary and pilot, Greene became the first woman to fly over the high Andes. According to Greene, her particular brand of feminism consisted of simply doing her job, which she saw as fulfilling the command of a higher power rather than proving that women could fly as well as men.[14] In the chapter titled "Woman in a Man's World" from her book, *Flying High*, Greene recounts her experiences during the 1950s while flying with the MAF in Sudan, a newly independent Islamic nation. She transported Christian missionaries, church leaders, or people needing help in emergencies, and once she piloted three Sudanese army officers who later invited her to lunch, during which one of the officers asked her to talk about her experience as a female pilot. Greene comments: "I found their treatment of me fascinating since I broke the stereotype of the role of Sudanese women, who are mostly restricted to their home and their children. To further break the stereotype, I was not married and I was doing man's work in a man's world."[15]

Greene explains why the officers accepted her, speculating that the men she'd just transported 175 miles to El Renk viewed her as a fellow professional and that perception transcended the usual gender barrier prevailing in Sudanese social situations. However, Greene notes that in other contexts she was treated more like the Sudanese women. She claims that her status as a pilot opened more doors for her in Sudanese society than her status as a missionary, and comments that despite her unusual occupation, she resisted becoming part of a "feminist crusade [unlike] ... many women in our society." Greene perceived her true liberation to be serving together with men "under Christ" and thought that when that happens, "the competition natural among the sexes can then give way to each serving the other."[16]

Even after WASP deactivation, Betty Pettitt Nicholas 44-W-7, who spoke Spanish and had been a maintenance test pilot at Napier Field in Dothan, Alabama, continued her career as an instructor for a Mexican student officer contingent located there.[17] Teresa James began working in her parents' flower shop, but she continued to fly and was commissioned as a major in the Air Force Active Reserve in 1950, retiring from service in 1976.[18] Hazel Raines, one of the Americans in the ATA, became a flight instructor in Brazil and later joined the United States Air Force; but this time, in contrast to her active flying experience as a WASP, her time was spent "flying" a desk in her role as a WAF recruiter.[19]

Soviet Union

Soviet pilot Olga Nikolaevna Yamshchikova, a pilot in the 586th Regiment, became a test pilot after the war and studied new advances in jet aircraft technology.[20] Zinaida Fedorovna Solomatina, also from the 586th Fighter Aviation Regiment, was hired by the Soviet Civil Aviation Fleet after the war, at first transporting people and medication in light aircraft, but eventually flying heavier twin-engined planes.[21] Antonina Skoblikova, an aircraft commander of the 125th Guards Bomber Aviation Regiment, says that, after the war, some of the women from that regiment returned to their jobs in civil aviation. She cited the following names: Sasha Yeremenko, Sasha Krivonogova, and Irina Osadze.[22]

In a 1988 interview on the occasion of the 50th anniversary of the *Rodina*'s flight, Grizodubova, who had been the pilot, deplored the fact that while young men seemed reluctant to train as military pilots, young women weren't even allowed to enter training; women could only join the paramilitary replacement for the *Osoaviakhim*, the DOSAAF, Voluntary Society for Assistance to the Army, Navy, Aviation and Defense Industry, which let them fly only sport aircraft.²³ When Cottam's *Women, War and Resistance* was published in 1998, women in Russia did not participate in either military or commercial flight.²⁴ When Strebe's book, *Flying for Her Country: The American and Soviet Women Military Pilots of World War II*, was published in 2007, there was only one Russian woman fighter pilot, Svetlana Protasova, and two female commercial pilots.²⁵

Great Britain

Pauline Gower was absolutely accurate in her prediction of the job situation women pilots would encounter at the arrival of peacetime. Former Women's Section pilots found it a challenge to pursue a career in aviation. Moggridge ends her chronicle of life in the ATA by describing her own postwar flying employment, which included delivering Israeli Spitfires to Burma in the 1950s. In 1953 Moggridge decided she wanted to be the first woman to set a British speed record by breaking the sound barrier before either Cochran (in America) or Auriol (in France) could seize the record; but when Moggridge petitioned them for help, the British Air Ministry rejected her request for financial assistance.²⁶ In the film *WASPs and Witches*, Moggridge recalls that the only postwar flying jobs available to her were either undesirable or dangerous because returning male pilots took all the others.²⁷ And, ironically, it was Diana Barnato Walker, a former member of the ATA Women's Section, who became the British woman who broke the sound barrier in a jet aircraft in 1963, flying faster than "the Jacquelines (Cochrane [sic] of the USA and Auriol of France)."²⁸

Veronica Volkersz joined the Royal Air Force Volunteer Reserve (RAFVR) but she found few flying jobs open to women, either commercially or in the reserves. In her 1956 memoir, *The Sky and I*, Volkersz anticipates only diminished opportunities for women pilots in commercial aviation: "We thought we had proved ourselves during the war, but some people have conveniently short memories. I firmly believe that women, even in this jet age, can hold down a flying job on equal terms with men. They are usually more careful pilots; they do their work more conscientiously; and what's more, they do not suffer from that preeminently male failing, the urge to show off."²⁹ She continues: "We had the excellent training and experience afforded by the ATA—in my own case, 1,200 hours on sixty different types of aircraft—and after the war, night and instrument flying practice in the Volunteer Reserve. Since, despite all these advantages, we have found it far from easy to get jobs, what chance has a girl of today, with no Civil Air Guard, no ATA, and not even a Volunteer Reserve in which she could gain experience and accumulate flying hours?"³⁰

Volkersz notes that in the mid 1950s, out of one-hundred ATA women, only seven were still flying commercially.[31] Volkersz thinks that the myth of male superiority is one element that contributes to the lack of opportunities for women pilots. She writes that "In my Chelsea flat I keep an illuminating file. It contains all the letters I have written to firms, all the advertisements I have answered, and all the replies I have received. Through the latter, echoing and re-echoing is a maddening refrain: 'It is not the policy of the company to employ women pilots.'"[32]

In her closing words, Volkersz finds some hope for the future of women aviators in the announcement that the Air League of the British Empire had drafted a postwar scheme to recruit and train civilian flyers — men and women — to fill gaps in the number of available pilots, but she was fairly certain that she would miss out on that opportunity because it would come too late.[33]

Rosemary Rees du Cros remembers the expectation many women pilots held of pursuing a commercial aviation career after the war, relying on their perceptions that barriers had broken down during their wartime service. She comments that many of them were disappointed because the old prejudices emerged anew along with the arrival of peace. After the war, du Cros earned her commercial license and started her own company, Air Taxi, flying a Proctor, an inexpensive war surplus airplane. She comments: "A great many little flying businesses like myself sprang up immediately after the war, and to start with they did quite well, but business fell away rapidly as the airlines got going and operated proper schedules to and fro, only charging passengers for the one journey. Gradually all the small ones I knew of folded up and if I had not got married and gone to live in the wilds I would have had to do the same anyway."[34] After making that statement, du Cros cites the common rationale underlying the commercial airlines' reluctance to hire women after the war when she speculates that the airlines "couldn't be blamed really" for not wanting to invest in training a woman who might then leave to get married and start a family.[35] The book *Women in Air Force Blue*, by Escott, notes that by the late 1980s, the RAF accepted WRAF women to train as pilots and navigators under the "same terms and conditions as men" with the exception of combat.[36]

Two chapters in Render's *No Place for a Lady* describe the return of Canadian women pilots from their wartime work either in England or in the United States, providing updates on the current activities of several of them. Vi Milstead instructed pilots; Elspeth Russell Burnett and her husband formed a flying service. Helen Harrison became a flying instructor, taking one of the few aviation jobs still open to women. Vera Strodl became a charter pilot in Sweden and then an instructor in England. Finding it difficult to get flying jobs in Canada, Strodl bought her own airplane in the 1960s so that she could spread the gospel more effectively. Marion Orr founded her own flying school, the first Canadian woman to own and operate one.[37]

WASP Militarization Achieved

Despite initial heartfelt assurances that WASP service would never be forgotten, the postwar reality for the women proved otherwise. When a small number of American women entered the U.S. Air Force as military pilots thirty years later, no one remembered that women had also flown military aircraft for the United States Army Air Corps in the 1940s. News reports referring to current female jet pilots as the "first women to fly military aircraft" so incensed a few former WASPs that they could no longer keep silent. The stage was set to reengage the fight to gain veteran status for the women who had flown for the Air Force as civilians but had been denied military status while they served. By the 1970s, the former women pilots were holding biennial reunions that began with the thirtieth anniversary of WASP formation.[38] They were poised to fight the perceptions of their young successors, who also assumed that they must surely be the first women to fly military aircraft, knowing nothing of the accomplishments of their predecessors.

The social climate of 1970s America was more favorable to accepting the WASPs as veterans than was the 1940s context. After the United States adopted an all-volunteer army after the end of the Vietnam War, women held more positions in the military than ever before because jobs previously unavailable to them had opened up.[39] The former WASPs had also become more politically sophisticated and they engaged in the legislative process to a far greater extent than they had during the earlier attempt. They studied documents and resolved not to make any mistakes. The women spoke freely about the work they had performed and argued that it constituted *military* service, an argument strengthened by the availability of previously sealed records from their service and by the powerful influence brought to bear on their male colleagues by female members of Congress. Some WASPs produced the honorable discharges they'd been issued.[40] In her book, *Dear Mother and Daddy*, Clark includes a copy of her own honorable discharge from the WASP.[41] Vera Williams, in *WASPs: Women Airforce Service Pilots of World War II*, claims that the timing for WASP militarization was propitious the second time around because America's new all-volunteer army was "wooing" women and it was therefore concerned that a second defeat of WASP legislation would have a negative impact on future recruitment.[42]

8—Conclusion

General Henry H. Arnold had died in 1950, but Arnold's son, Colonel W. Bruce Arnold, in an effort to "finish his father's work," championed HR 8701, the GI Bill Improvement Act, which was sponsored by Senator Barry Goldwater of Arizona and introduced to Congress in March 1977. The bill included an amendment giving veteran status to the WASPs. Senator Goldwater had himself flown ferrying missions in World War II while based at Newcastle AAB. He had worked alongside female ferry pilots and had been impressed by their performance. Goldwater was a strong advocate for the new WASP legislation.[43] In addition, HR 8701 was cosponsored by all the female members of Congress. Colonel Bruce Arnold testified, stressing the equivalence of WASP service to the work that had been performed by men. He pointed out that male pilots had been recognized for their contributions while WASPs had not been, even in cases when they were killed or injured while performing identical duty.[44]

In her effort to validate WASP service, Byrd Granger, a member of the first class, gathered documentary evidence proving that the American women pilots had indeed performed military service. Granger compiled these raw materials into a report, *Evidences of Military Service by Women Airforce Service Pilots in WW II*.[45] Her book, *On Final Approach*, is the highly detailed account based on her research, essentially a day-by-day narration about the formation, management and outcome of the WASP program. Published posthumously in 1991, the book lists what Granger considered the persistent misconceptions about the women pilots, as follows: (1) that the program ended because Arnold grew tired of negotiating conflicts between Love and Cochran, (2) that more WASPs flew ferry missions than in any other command, and (3) that ferrying was the most important work the pilots did. Granger claims that all three statements contain some truth, but they are basically untrue. Granger also identifies misinformation concerning their numbers, flight hours logged, and the social background of the first group of WAFS.[46]

Dora Dougherty Strother 43-W-3 supplied a history of the WASPs for Congress that also proved that the women's service was military in every way.[47] Another piece of evidence was the photograph of Adela Scharr in uniform among a group of similarly attired male pilots.[48] Bee Falk Haydu commented that graduates from the training program were assigned to either the Ferrying Command or the Training Command. One photo in her book, *They Also Served: American Women in World War II,* shows Haydu, then WASP president, standing with Arnold and Goldwater prior to the senate hearing on the second WASP bill.[49]

The timing of the bid for WASP militarization, so problematic in 1944, was practically perfect more than thirty years later. On November 3, 1977, the U.S. House of Representatives passed the bill, as did the U.S. Senate the next day. President Jimmy Carter signed WASP militarization (PL 95-202) into law on Thanksgiving Day, November 23, 1977.[50] The new law gave veteran status to all the women listed as WASP participants, including those who trained in the program but did not actually serve. Interviewed later, former WAFS Barbara Erickson London expressed displeasure with that inclusive decision on the grounds that veteran status was not automatically granted to any other service personnel who had failed to graduate from a training program.[51]

Despite their new status as veterans, the women, or sometimes their heirs, still needed persistence when claiming their rightful benefits, as in the case of former WASP Irene Englund's daughter, who was at first denied the right to bury her mother's remains at Arlington National Cemetery in 2002, despite the fact that Englund was now officially a veteran. Her daughter protested the cemetery's denial in a letter to the *Washington Post*, prompting a review of the rules by assistant secretary of the army Reginald Brown. The army reconsidered its decision and allowed Englund an Arlington burial with "standard honors."[52] The film *Women Combat Pilots* shows scenes from Englund's burial at Arlington National Cemetery with full military honors in 2002, but it acknowledges that there had been "omissions in the past."[53]

Questions

Even now, new evidence from wartime incidents continues to appear. Sometimes that evidence sheds light on events that could not be explained at the time. However, mysteries remain and there are some questions that still need deeper investigation.

What Happened to Amy Johnson?

In early January 1941, in uncertain weather conditions, Amy Johnson set out in an Airspeed Oxford from the airfield at Squires Gate, en route to Kidlington, 150 miles to the south.[54] Flying above the cloud layer, she was apparently unable to locate an opening through which she could see the ground. Johnson had adequate fuel for several hours' flying time, but after a while she probably parachuted from her airplane over the Thames River estuary in icy conditions. Rescuers on a ship in the vicinity, having located her, were not successful in retrieving her from the water, and she was lost, her body never recovered. In a 2003 biography, Gillies notes that Johnson's tiny Airspeed Oxford has not yet been salvaged from the shallow water in which it sank, despite a 2002 search.⁵⁵ According to Gillies, recovery of the airplane would be difficult because of the constantly shifting currents in that section of the river.

Luff, another Johnson biographer, presents every fact at his disposal pertaining to her fate and he considers each rumor that arose in turn. There was speculation that Johnson was carrying a passenger and that she was involved in a secret mission, that she did manage to parachute from her airplane, or that she landed and stayed with the plane until it sank. Some believed that her airplane was shot down by enemy fire; others think it was brought down by "friendly" fire. Luff concludes that Johnson was the victim of a nearby ship's antiaircraft fire. He writes that no official inquiry into Johnson's death was ever conducted and notes that efforts by her family members to solve the mystery of her final flight have been unproductive.⁵⁶

In *ATA Girl: Memoirs of a Wartime Ferry Pilot*, du Cros describes her acquaintance with Johnson and speculates on how she thought the famous aviator met her death. She thinks Johnson probably ran out of fuel while flying in thick clouds over the estuary and that she must have desperately tried to find a place to land.⁵⁷ The film *WASPs and Witches* also claims running out of fuel to be the reason for Johnson's crash. There is footage of a man who testifies that he was the one who had mistakenly shot down Johnson's plane over the Thames, but he was told at the time by his superiors to keep quiet about it.⁵⁸

What Was Lily Litvyak's Fate?

On August 1, 1943, Lily Litvyak departed on her fourth combat mission of the day. The famous fighter pilot, so skilled that she was allowed the privilege of searching out targets of opportunity as a "free hunter," never returned.⁵⁹ The question remains: was the famous fighter pilot killed in the battle between nine Soviet and forty German aircraft, or did she survive the battle but become a prisoner of war? For years, searches for Litvyak and her airplane were unsuccessful. Rumors spread concerning her ultimate fate. In 1988, the remains of a small female with a bullet hole in her skull were discovered in a location other than the place where her fellow pilots thought she must have gone down. By deducing the pilot's identity through the records of missing pilots in the area, the remains were identified as Litvyak's.⁶⁰ However, a book published in 2004 disputes this conclusion. It claims that a Soviet prisoner of war noticed the diminutive pilot in his camp, fueling speculation that she may have been captured and even might have used the circumstance as an opportunity to escape the Soviet Union.⁶¹

What Caused the Crashes?

In the chapter titled "We Lose One" in *Sisters in the Sky*, Scharr describes the mid-air collision of two airplanes over the state of Texas in March 1943 resulting in the death of WAFS Cornelia Fort, the first of the thirty-eight American women who died in service as WAFS or WASPs. Scharr claims that Lt. Frank W. Stamme, Jr., was at fault, perpetrating a "joke [that] had become harassment."[62] Simbeck, in a biography of Fort, reconciles two contradictory accounts of the accident which took her life. He challenges the commonly accepted version of Fort's last moments that claim she was the innocent victim of male horseplay. Evaluating accounts from eyewitnesses to the collision, Simbeck concludes that inattention, rather than hostility or horseplay, was the reason behind the crash. The two airplanes were violating WAFS policy to begin with by "flying formation," and they were far too close to each other for safety. Simbeck interviewed three of the pilots who were flying in formation with Fort during her last flight, as well as an eyewitness on the ground and the person who investigated the accident. One of the pilots contacted by Simbeck told him, "I've been waiting over fifty years for this call."[63]

Ambiguous testimonies surround WASP Hazel Lee's collision with another P-63 on her final approach to East Base at Great Falls, Montana, on November 23, 1944.[64] She was killed delivering a P-63 for the Russian Lend-Lease program. On final approach she pulled her airplane up (as directed by the control tower) into another airplane, which had also been told to pull up. Lee's life and death became a matter of deep importance to Kay Gott, her biographer, who was dogged in her attempt to ferret out information about Lee's fatal crash. Her investigation of the accident turned up eyewitness accounts that contradicted the official accident report. Gott implies that the speed with which the accident investigation was accomplished might hint at an attempt to avert further inquiry. In *Hazel Ah Ying Lee: Women Airforce Service Pilot, World War II, a Portrait*, Gott presents a copy of the *Army Air Forces Report of Major Accident*, a "confidential" account of the accident—marked with arrows, underlinings and the word "Blame" attached to one paragraph—in which the reason cited for Lee's fatal accident was "failure of [the other pilot] Russell's radio, the failure of the Wasp pilot to make a 'check call' on 'Base Leg' before landing, and failure of either pilot to heed the red light warning given by the Tower [illegible word]."[65]

Gott presents two brief preliminary accident reports, but she notes that Lee's flight logs were unobtainable. Gott, who was also ferrying a P-63 to Great Falls on the day of the accident, provides entries from her own logbook and those of Helen Schaefer, Isabel Madison, and Gretchen Gorman, all

WASPs who were delivering airplanes that day.⁶⁶ A preliminary accident report reproduced by Gott concludes that a "P-63 turning from short base and letting down at same time" and "PROBABLE tower error" were likely causes of the collision.⁶⁷

WAFS Evelyn Sharp was killed when the P-38 she was ferrying developed engine trouble on takeoff from a Pennsylvania airfield. The plane could not maintain altitude and it crashed, throwing Sharp from the cockpit and killing her instantly. Bartels, in the chapter notes for her book, *Sharpie: The Life Story of Evelyn Sharp, Nebraska's Aviatrix,* writes that despite her efforts, she had no success obtaining access to the official records of Sharp's accident, noting that "much of the P-38, 3 Apr. 1944, accident description is censored." Bartels invoked the Freedom of Information Act and challenged the ruling that denied her access to the records; even so, she could not obtain information about the airplane during the time it was in New Cumberland, Pennsylvania, on 2–3 April 1944.⁶⁸

Were There Japanese Women Pilots?

What role did female aviators play for Japan? In *Battle Cries and Lullabies,* De Pauw refers to sources citing Japanese women as combatants, especially in the last years of the war. Female bodies were found in the wreckage of downed Japanese bombers, leading to speculation that the women might have been bombardiers or wireless operators.⁶⁹ It seems plausible that the abilities of women as pilots or other kinds of aviation personnel would have been tapped by Japan, especially near the end of the war.

Luck or Miracles

Diana Barnato Walker, a self-confident woman of privilege, learned to fly at England's Brooklands Flying Club, which had been established along with the Civil Air Guard as the British response to Germany's Youth Flying Club movement. Walker first soloed in 1938, but not before she was warned about flying by a man deeply scarred from an airplane crash before she was about to takeoff. Throughout her autobiography, *Spreading My Wings*, Walker harks back to the man's warning; although she ignored him before her first solo flight, she credits his words for saving her life more than once. Walker was unusually lucky during the war or, as she claims, protected by a guardian angel, because the several close calls she experienced left her uninjured in circumstances identical to those that had killed others. Once, Walker managed to land her Spitfire in a grassy airfield, wet with puddles and surrounded by trees; but the low clouds surrounding her had already taken their toll on other pilots at the airfield, leaving several aircraft smashed and two pilots dead.[70]

On a cross-country flight from Long Beach, California, to Phoenix, Arizona, Geri Lamphere Nyman, 43-W-1, experienced what she considered a "miracle" landing adjacent to a Goodyear runway in Luke, Arizona. A dust storm prevented Nyman from flying to her original destination. Searching for an alternative, she noticed a likely airfield; but upon inspection from above, she realized that she'd need to approach the field close and low. Nyman says she then felt "something she couldn't explain" taking over the airplane's controls, and — against her instinct — forcing her to land just beyond the end of the runway. This action prevented Nyman from flipping her airplane over in the soft newly asphalted runway pavement. Once she was on the ground, onlookers rushed up to her and one exclaimed, "Lady, you've got some kind of angel riding on your shoulder."[71]

Paradoxes

Babington Smith, in her biography of Amy Johnson, cites her letters and others' recollections proving that Johnson's brief time ferrying in the ATA was the happiest and most fulfilling period of her life. By then, the famous aviator had reached a personal conclusion that her constant quest for new aviation records to break was a hollow pursuit, lacking the satisfaction afforded by the critically important work she did in the Air Transport Auxiliary.[72] In describing her own service with the ATA, Leveaux writes, "It was a once-in-a-lifetime experience, and we felt both fortunate and proud." She also acknowledges how difficult it was to return to a "woman's role" in 1946.[73] Barbara Erickson London refers to her experience ferrying aircraft with her fellow WAFS as the "best time of our lives."[74]

Soviet women pilots flew in all kinds of aerial combat, their sex never exempting them from the same responsibilities assigned to males; but it was the women's very femininity that, according to one woman pilot, gave her regiment an edge in combat.[75] Raisa Yermolayevna Aronova flew in the 588th Night Bomber Aviation Regiment, later the 46th Guards Regiment. After the war, Aronova stated that she believed that military service was inappropriate for women and further, that the situation in the Soviet Union during the Great Patriotic War was exceptional. However, Aronova credited the success of her own regiment to the fact that it was all-female throughout the war, and therefore retained its feminine characteristics.[76]

Another of the Soviet "night witches," Yevghenia Andreyevna Zhigulenko, was in Moscow digging trenches next to her friend, Tanya, while German troops advanced on the city in the autumn of 1941. Suddenly planes strafed the area where they were working. When the attack was over, Zhigulenko lifted her head, but her friend lay dead. Zhigulenko describes that moment: "Understand, we had just been conversing, laughing, discussing this, planning that, and suddenly — she is no more. Suddenly, this human being ceased to exist! At that moment, I understood finally what war was."[77] Night bomber Zhenya Rudneva entered these words in her diary on December 2, 1942, summing up a wartime experience shared by many: "I thought about it recently, and a silly idea, a complete paradox, came into my head: after all, the war is on, there is so much horror and spilled blood all around, while for me, I am sure, this is the best time of my life."[78]

The Legacy

Were the military pilots who happened to be female only a wartime aberration? Or did the women's participation in a male dominated activity open doors of opportunity for them afterwards, even if only slightly? The women pilots themselves can cite countless instances of male preconceptions that were altered when the men had an opportunity to observe capable female aviators at work. There is no documentary evidence proving that these pilots fell short of performance standards during their service. On the contrary, "experiments," like the Women's Section of ATA in Great Britain or the WASP in the United States, concluded successfully. Yet, except in rare individual cases, the air forces did not retain women pilots after the war ended, and governments throughout the world encouraged the women to abandon their nontraditional and highly temporary work and resume women's typical roles.

The film *Women Combat Pilots* states that 5300 American women served as pilots in the First Gulf War in the early 1990s, but they did not fly in combat. In the film, a male pilot describes the strength of institutional resistance within the U.S. military to women pilots like Kara Hultgreen, who discovered that, as a female, she would be restricted to flying transport or instructing others because she could not engage in combat. Hultgreen lobbied against women pilots' second class status during the Gulf War, and she urged Congress to give them equal rights.[79] America's 1949 combat exclusion law was repealed in 1994, and Hultgreen then trained to fly Navy F-14s in San Diego, California, becoming the first woman in America to qualify as a combat pilot. Hultgreen was stationed on the USS *Abraham Lincoln*, which immediately became known by a new nickname, the "*Babe-raham Lincoln.*" The film interviews Hultgreen's mother, who gives the simple reason why, at the time, women in the navy were assigned to a ship as a group, giving them priority over male pilots. The women had to serve on the only aircraft carrier that was equipped for them. The elder Hultgreen speculates that this accommodation fueled jealousy among the male pilots, feelings that were probably justified, she comments. Men claimed "political correctness" was behind the navy's decision. Hultgreen was killed while making a nighttime landing onto the aircraft carrier. She ejected into the water before her plane crashed. Some thought that her inability to fly caused her death, and that conclusion put the abilities of all women pilots into question. The navy's investigation determined the cause of the crash to be engine failure.[80]

One woman pilot, interviewed for the film, notes that aircraft carriers were "male clubs" and male pilots feared that standards would fall with the arrival of female pilots. Nevertheless, one male pilot comments that the *airplanes* don't care who's flying them and men and women both face risks and dangers in wartime. He notes that the characteristics distinguishing pilots from other personnel are more important than the differences between male and female. All pilots are type–A personalities, he says. He goes on to note that the dangers of capture are equivalent for men and women because both sexes can be raped.[81] This film also features a black female pilot, and it optimistically concludes with the statement that women are now "fully integrated" into American military aviation because the myths about women's inferior capabilities were exploded as soon as the men saw that women could fly equally well.

Unaware of Helen Richey's stunning but short career with Pennsylvania's Central Airlines in the mid–1930s, Bonnie Tiburzi, in *Takeoff!*, claims that she was America's first woman pilot for a commercial airline. Her failure to give credit to her long-forgotten predecessor notwithstanding, Tiburzi's accomplishment is significant. Tiburzi preferred working for a large commercial airline because it meant better job security than could be provided by smaller aviation companies. When Tiburzi was hired, there were very few female pilots for the major airlines; she cites four in the United States in 1973, and one in Canada.[82] In closing, she acknowledges the legacy of the American WASPs, noting that Barbara Erickson London's daughter, Terry, was employed as a pilot by Western Airlines.[83]

When interviewed for the film *Women of Courage*, Byrd Howell Granger said that participants in the WASP program didn't understand, when they served, why any particular fuss should be made over them. As she observes in the film's early minutes, "I've never been in an aircraft yet that asked what sex I was.... [I]f you knew how to fly you flew!"[84]

How They Saw Themselves

All of the women who were pilots in the Second World War broke new ground. They challenged—and sometimes changed—commonly accepted assumptions about the capabilities of women and their roles, particularly in wartime. The women were obliged to prove their capability over and over. When given the opportunity, they could—and often did—change the preconceptions of others. Because the fact that women were flying military airplanes wasn't widely publicized, even within the air forces, female pilots were often mistaken for assistants, nurses, or prostitutes. When they were first assigned to air bases they often faced hostility from their fellow pilots and sometimes from their commanders. Occasionally the women's own perceptions were inaccurate because they too were not immune to an unthinking acceptance of common stereotypes.

Photographs of the American, British, and Soviet women pilots taken while they served reflect the contrasts in their different social contexts. Hairstyles and makeup, or lack of makeup, constitute the most apparent differences. The American and British women adopted the era's longish, permanently waved, side-parted hairstyle, but most of the Soviet women, including Raskova, wear their hair combed flat, often parted in the center, with long braids wrapping across the crown. Their photographs show no evidence of lipstick or other makeup. Appearances could fool the women themselves. Jean Hascall Cole, in *Women Pilots of World War II,* tells a story about her 44-W-2 classmate, Nellie Henderson, who was dining in the officers' mess while at officer candidate school. Henderson, smartly attired in her WASP uniform with its silver wings, blithely chattered with her tablemates about the risks of her work towing targets, and she ignored the man next to her who kept poking her in her side. When the woman who'd been seated across the table left, Henderson queried the others: "Who was that awful looking thing over there?"[85] The woman turned out to be a Soviet war ace who had downed seventeen airplanes. The embarrassed Henderson could only hope that the pilot knew no English!

In their recollections, Soviet women in Raskova's regiments often refer to each other as "fine fellows."[86] They call each other "sisters," and claim that their comrades were as close to them as family. Polina Gelman, who served in the 46th Guards Night Bomber Regiment, says that the women's regiments were more disciplined than the men's; and because the women were volunteers, they served because they wanted to defend their homeland and therefore they were imaginative and passionate warriors.[87]

WAFS Teresa James describes the camaraderie that developed among the women who flew in the United States: "The women like you were sisters, flying around the country, and then coming back and chatting about it — just like a big family.... You became part of each other because you could talk about what happened, the scary moments, and bolster each other. You became very close."[88] However, quite a few personal testimonies indicate that the WAFS and the WFTD never really connected, even after they merged into the WASP. It is therefore naïve to assume that all the women who flew for the United States Air Force during the Second World War share an implicit bond. While they served, more factors conspired to separate than to unite them. Stationed at airbases scattered throughout the country, the women, whose job was ferrying aircraft, constantly moved from place to place. The two groups began their service in separate organizations, or, as WASPs, in one of the eighteen individual classes, so they were usually acquainted only with their classmates and perhaps a few others. They tended to develop deep loyalties to their original group, and to one or the other of their two leaders.

The pilots' unusual occupation altered some of their perceptions. For instance, Du Cros writes in her memoir that news of an ATA pilot's death from illness was greeted with deeper consternation than news that he or she had been killed while flying. The potential of losing one's life in an aircraft during wartime was an accepted risk that all the ferry pilots took, but news of a premature death due to illness could cause them genuine shock.[89] In contemplating the likelihood of being killed while ferrying for the ATA, du Cros writes that she would have been more dismayed by the thought that her fellow pilots would think her an incompetent pilot than by the idea of losing her life in an accident.[90]

Part 2

Biographies

9

The Leaders' Stories

An essential element in the women pilots' participation in World War II was the influence of charismatic and powerful leaders. This chapter discusses the four women without whom other women pilots would probably not have flown military aircraft in the war — Cochran, Gower, Raskova, and Love. Their biographies are arranged chronologically by their birthdates. Chapter 10 has the life stories of a few of the women pilots from each country.

Jacqueline Cochran (1906–1980)

Although she claimed to be an orphan who knew neither her real parents' identity nor her precise birthdate, the facts revealed by a recent biography of Jacqueline Cochran contradict her widely accepted life story through a well-documented investigation into the famous aviator's early years. Born in Muscogee, Alabama, on May 11, 1906, to Ira and Mollie Pittman, the wealthy beautician and aviator began her life as a little girl named Bessie in an impoverished family that traveled throughout logging camps and mill towns in Florida.[1] In *Jackie Cochran: Pilot in the Fastest Lane*, Rich writes that Cochran claimed one day to have overheard a conversation in which Mollie remarked that six-year-old Bessie was "not one of the family," and those words became the foundation for her tale of having been orphaned at age four.

Having completed only two years of grade school, Cochran struggled all her life with written directions and tests, and she often requested oral exams as a substitute. Her native intelligence combined with those accommodations gave her the means to succeed. Ann Wood, who served as Cochran's assistant while she was in England, would later recount her employer's habit of privately asking for the definitions of words she hadn't understood in conversation, and then consciously using those new words in her daily speech until they became part of her vocabulary.[2]

While she was still living in Florida, Cochran worked briefly as a nurse, but concluded that there was no desirable future in tending to the poor, so she decided to become a beautician. Filling in the omissions in Cochran's accounts, Rich discloses that at age fourteen Bessie became pregnant. She married the twenty-year-old father, whose surname was "Cochran." In 1921 she gave birth to a son who would die in a fire four years later. By then, the child was in the care of his grandparents in Florida, his mother having moved to Montgomery, Alabama, in pursuit of a beauty operator's career. Cochran and her husband divorced. Although Mollie Pittman lived to see her daughter become famous, Cochran's father, Ira, died in 1928. In 1929, when Bessie arrived in New York City with the intent of re-creating herself in a place far away from the squalor of her childhood, she assumed the name "Jacqueline Cochran," using a surname she claimed to have chosen from a phone book. Cochran characteristically began her job search in her new home by heading straight to the top, and she first arranged a meeting with Charles of the Ritz, a bid for employment that was rejected; then she set her sights on a competitor, the famous Antoine, who was impressed by the young woman's spunk and immediately hired her.

While spending the winter in Miami Beach with a client, Cochran met her future husband, Floyd B. Odlum, one of America's wealthiest and most influential businessmen. She was already a successful beautician, but Cochran's marriage gave her even more opportunities to pursue her dreams. Before they married, Odlum challenged her to get her pilot's license so she could manage her expanding cosmetics business more easily. Reluctant at first to reach so far outside her normal sphere, Cochran rose to Odlum's challenge and obtained her pilot's license in half the time her future husband predicted she'd take. It was in taking her very first flight that she claimed to have discovered her life's consuming passion.

Amelia Earhart was among Jacqueline Cochran's very few close friends. The two women shared a bond of psychic ability, in addition to their devotion to aviation. Concerned that the tiny island in the Pacific Ocean which Earhart had chosen for a refueling stop would be too difficult to locate from the air, Cochran urged her famous colleague to abandon the attempt for a round-the-world record. After Earhart's disappearance in 1937, Cochran, who had previously entered air races but, for various reasons, had never completed them, won the Bendix Transcontinental Air Race in 1938, and was named "First Lady of the Air Lanes" in the *New York Times Sunday Magazine*.[3]

In the interwar years, she flew in a number of air races and set new speed and altitude records. However, she challenged the practice of maintaining separate aviation records for men and women. She scorned so-called women's races and she had no interest in achieving a "woman's record." Replying to an article in *Air Force* magazine, which mistakenly listed her as the holder of *women's* records, Cochran retorted, "A record is a record, whether made by man or woman and there would not be two."[4] True to herself, but contrary to the era's conventions, Cochran insisted that her aviation records be recorded in her own name, "Miss Jacqueline Cochran," *not* "Mrs. Jacqueline Cochran Odlum." Cochran's relationship to other women and to women's causes was often thorny. At first she refused to join the Ninety-Nines, the organization of women pilots formed in 1929 by Amelia Earhart, its name signifying the number of the club's original members. However, Cochran later accepted an offer from the 99s to make her their president after giving a speech that promoted women pilots in the war effort.[5]

Some of her contemporaries appreciated Cochran's forceful personality, but she was actively disliked by others. People's reactions were strong on both sides throughout her life. Adela Scharr, while attending OCS in Florida, describes her impression of a visit by Jacqueline Cochran. It was while the WASP legislation was pending and just after columnist Drew Pearson's unflattering piece on the women pilots had appeared. Scharr, an original WAFS whose loyalty to Nancy Love is obvious, writes that she was impressed by the commander of the WASPs after seeing her in person.[6]

Cochran's marriage to Floyd Odlum was a happy one and their personalities complemented each other. With Odlum, who transcribed her spoken words and whom she dubbed her "wing man," Cochran wrote an autobiography, *The Stars at Noon*, which was published in 1954. Odlum is conspicuously absent from Cochran's book, which blends her personal philosophy and life story. The autobiography is entirely consistent with the persona Cochran so carefully crafted after she left her impoverished birthplace behind and embarked on her journey to bring her soaring dreams to reality. Authoring a book presented one more opportunity to change the story of her humble origins and create the woman she wished to be.

Jacqueline Cochran was convinced that women pilots could play an important role in wartime and she wanted to initiate a training program for them in all types of utilitarian aviation work, short of combat. Before America entered the Second World War, Cochran assembled a small contingent of women to join the Women's Section of the British ATA whose purpose would be to determine whether women were capable of handling military aircraft, not just to assist the British. The "experiment" was successful, so Cochran then established a training program for women pilots in the United States and became the director of the Women Airforce Service Pilots, until the group's disbandment in 1944. General Henry H. Arnold then charged her with the task of writing a final report on the WASP program.[7]

In August 1945, Cochran was the only woman, among fourteen civilians, to be presented America's Distinguished Service Medal.⁸ She became a lobbyist for independent status for the United States Air Force. In fact it was she who urged Arnold to act quickly on his request for independence from the army. Cochran writes that she traveled across the country for a year making speeches on behalf of a separate air force. She joked at the time that the army would get sick of her and let the air force go just to get rid of her.⁹

After the war ended, Cochran continued flying privately, managing her beauty business and entering air races. In 1951, she campaigned for Dwight D. Eisenhower's presidential bid. In 1953, Cochran and friend Chuck Yeager flew separate F-86 Sabre jets; they simultaneously broke the sound barrier, an event chronicled in Yeager's introduction to *The Stars at Noon*.¹⁰

Jacqueline Cochran died in Indio, California, on August 9, 1980, and was buried, as she had requested, with the long cherished and eventually reclaimed doll that she had won in a contest during her childhood but was forced to give away.

Jacqueline Cochran in her WASP uniform. She selected a designer and manufacturer and chose the Santiago blue color in anticipation of the WASP joining the U.S. Air Forces (The Woman's Collection, Texas Woman's University).

Pauline Mary de Peauly Gower (1910–1947)

Pauline Gower was born on July 22, 1910, in Kent, England. Her mother was a musician. Her father, a solicitor, would later become a member of Parliament and the mayor of Tunbridge Wells. Consistent with her social class, Gower was destined to be presented to society and marry comfortably. She would never have needed to concern herself with earning a living. Her father, Robert Gower, although not a Catholic, sought the best education possible for his two daughters, an unusual aspiration for female children then. So Pauline attended the Tunbridge Wells Sacred Heart Convent School and later a finishing school in France, which she left early in order to pursue her career in aviation. Gower considered the prospect of a socialite's life boring. She took her first airplane flight at age eighteen. Without her father's approval, and consequently his financial support, she funded her own flying lessons by giving lectures and playing violin recitals. In the summer of 1931, Gower earned her pilot's B license, which gave her the certification necessary to carry passengers. At twenty years old, Pauline Gower was only the third—and youngest—woman in the world to earn a B license.[11]

Pauline Gower formed an air-taxi and joyriding service, Air Trips, with her friend and business partner, Dorothy Spicer. On the occasion of her twenty-first birthday, Gower's father, having adjusted to the inevitability of his daughter's interest in aviation, presented her with an airplane, a secondhand Spartan two-seater. Later, Gower admitted to some manipulation of her parents in the quest to obtain her own airplane, having regaled them with stories of the forced landings she'd made in unsafe aircraft when the more likely reason was her own inexperience in flying. Gower's youthful ploy—combined with her parents' ignorance of aviation—did get her a more airworthy craft. She was determined to support herself by flying, so she was delighted when she and Spicer earned the sum of three pounds on their first day in business. Eventually Gower and Spicer incorporated Air Trips, convinced that they could make a living through aviation. The two women worked in air circuses for several years during the early 1930s. A few close calls in flight spurred them to hone their skills, and as a result, Gower learned both formation and cross-country flying. She chose not to participate in air races because of the expense. By 1934, Gower had made 10,000 passenger flights.[12] She joined Britain's National League of Pilots that year and obtained her instrument flying certificate and her navigator's certificate in 1935.

The summer of 1936 brought very bad flying weather to Britain. After the year was over, the two partners, both weary of the dangerous and uncertain life of air circus performers, and each having suffered family losses, amicably parted ways. Although it is chiefly concerned with the adventures of two women joyriders, the closing chapter of Gower's *Women with Wings*, which she wrote just after Air Trips ceased operations, encourages her contemporaries to take flight lessons and gives concrete advice on getting started. Gower lists flying schools by name, recommends a medical exam, and suggests wearing breeches or trousers, attire that was unusual for women at the time. She describes the tests necessary to qualify for an A or B license, and advises women to take their lessons seriously, claiming that no one is born a pilot, but achieves proficiency only through hard work.[13]

Pauline Gower was surprised that she and Spicer were the only two females pursuing a career in commercial aviation during their Air Trips years. In her book, she notes that her first passenger actually was a woman, a person Gower commends for having the courage to fly with a pilot of such limited experience! Although she knew that women were capable of becoming pilots, Gower acknowledged that marriage might end their aviation careers, especially if they preferred to be close to home. She gave three reasons for her interest in aviation: her love for flying, her determination to work in a paying career, and her conviction that aviation was the profession of the future.

Gower shared her father's passion for animals and she also contributed time to the RSPCA. An engaging writer, she produced several articles and short stories during the 1930s, including two volumes of whimsical poetry, *Piffling Poems for Pilots* and *More Piffling Poems*, that show her humor and playfulness.[14]

During the Second World War, Pauline Gower was instrumental in creating the Women's Section of the Air Transport Auxiliary, a group of civilian pilots whose purpose was to ferry aircraft from factories to air bases in England in support of the RAF. She was promoted to the highest rank possible in the ATA short of the rank of leader which was the position held by Commodore Gerard d'Erlanger — director of women personnel.[15] When Gower was awarded the Member of the Order of the British Empire in February 1942, she responded by crediting the performance of the women of the ATA for her success, claiming that she didn't know why she had been chosen for the honor. Gower helped effect a change to established British treasury policy, which dictated that women could earn only eighty percent of a man's salary for the same work; by May 1943 ATA women were paid the same as ATA men.[16] During her wartime service, Gower was appointed to the board of BOAC, the administrative agency of the ATA, which doubled her workload. She was frequently under doctor's orders to rest because she suffered periods of physical decline due to residual weakness from a serious illness contracted during her convent school days.

Pauline Gower met Bill Fahie in early 1944 and they were married in June 1945.[17] She died on March 3, 1947, after giving birth to twin sons, one of whom, Michael, gathered her personal documents and her family and colleagues' individual recollections to compile a biography of his mother.

Marina Mikhaylovna Malinina Raskova (1912–1943)

Marina Raskova was born in Moscow on March 28, 1912. Her mother was a schoolteacher, her father a voice instructor. Because Raskova had musical talent, she aspired to a career in the opera, but changed her course at the age of fifteen and set out on a path that would end in a career in aviation. Suffering from extreme stress and an ear infection, she needed to decide whether she would pursue music or science. Raskova opted to discontinue her study of music in favor of learning chemistry, thinking that a scientific field would be more likely to yield a promising career. In 1929 she married an engineer, Sergey Raskov, and gave birth to a daughter, Tanya, in 1930. Raskova and her husband divorced in 1935.[18]

As a draftswoman at the Aero Navigation Laboratory of the Air Force Academy in 1931, she studied the theory of navigation, becoming the first woman navigator in the Soviet Union, and later she learned to fly.[19] She became an instructor of older male officers who had been recalled to active duty from the reserves, and she also flew in air shows and entered air races. Working with a few other pilots, she set long distance aviation records and became a career officer, a lieutenant, in the Soviet military. She and two other women, Valentina Grizodubova and Polina Osipenko, set a world distance record — 6,450 km — when they flew nonstop from Moscow to the Soviet Far East in 1938.[20] Realizing that their fuel was too low to achieve their designated landing point, Grizodubova, the pilot, ordered the navigator, Raskova, to bail out of the airplane because the pilot intended to bring it down in the tundra. The story of the women's flight, and later of Raskova's journey through the Siberian wilderness before their rescue, was of intense interest to the Russian people, who eagerly followed radio reports on the women's progress. The three aviators were honored with a national celebration and each was awarded the Gold Star and the title Hero of the Soviet Union, the only women so honored before the Second World War.[21] During that period, Raskova, while recovering from her ten-day trek after the flight, responded to fan mail and wrote an autobiography, *Zapiski Shturmana* (*Notes of a Navigator*), which was published in 1939.

According to Pennington's "Raskova, Marina Mikhailovna Malinina" in the encyclopedia *Amazons to Fighter Pilots: A Dictionary of Military Women*, Marina Raskova became a member of the Communist Party in 1940 and attended the M.V. Frunze Military Academy. After Germany invaded the Soviet Union in June 1941, she used her considerable influence to lobby for the participation of women pilots in the war effort. She obtained permission from Joseph Stalin to form an all-female flight group with herself as their leader.

In her personal memoir, Militsiia A. Kazarinova, who served as chief of staff of Raskova's 587th Dive Bomber Aviation Regiment, recalls that Raskova was a good listener and an inspirational leader who could solicit the best performance from the "girls." In the mess hall, Major Raskova ignored the separate room where her own dinner had been served, choosing instead to eat soup with the regiment's technical personnel, and afterwards telling them that they need not report to her about the food because she'd tasted it herself. She then expressed her hope that they would do better the next time.[22] Raskova's speaking style has been characterized as respectful and charming, and therefore "disarming." The women under her command remember that she sang frequently and played the piano, true to her musical bent. Her faith in her regimental commanders was evident in the way she patiently worked with them to improve their performance. Raskova, who had a powerful influence on the Soviet leadership, was also loved and respected by the largely civilian volunteers she transformed into an effective fighting unit.[23]

Although well trained as a navigator, Marina Raskova was not a highly experienced pilot. She was killed while leading a flight of three aircraft from an intermediate airfield to her own regiment's location at the Stalingrad Front. She had ordered the two other pilots not to leave the flight formation except in an emergency. Senior Lieutenant Galina Tenuyeva Lomanova, the pilot of one of the three Pe-2s, describes her experience that day after the weather closed in so severely that she could no longer see the wings of her aircraft. Suddenly her navigator yelled, "There is the earth!" and almost immediately they belly landed. All three airplanes in the formation made emergency landings that day, but only Raskova's plane, which hit an embankment, had fatalities. Her three male crew members were also killed. Lomanova recalls that Raskova made the decision to fly, even as the weather deteriorated, because she was anxious about her regiment; but Lomanova thinks that her commander showed poor judgment in her determination to get to her destination no matter what.[24]

Raskova's death on January 4, 1943, left the women in her regiment devastated. They had already pledged to her that they would achieve the Guards distinction, and after her death they lived up to their oath, according to Kazarinova.[25] The 587th Dive Bomber Aviation Regiment so distinguished itself in battle that it was renamed the 125th M.M. Raskova Borisov Guards Bomber Aviation Regiment in September 1943. Marina Raskova's ashes were interred in the Kremlin. Several streets, schools, and a large ship were named for her and monuments were erected in her honor.[26]

Nancy Harkness Love (1914–1976)

Nancy Love was born into a well-to-do physician's family on February 14, 1914, in Houghton, Michigan. She spent a year at Milton Academy in Massachusetts. Love's interest in aviation, never impacted by the high cost of flight lessons, started when she took her first flight at age sixteen in 1930. During flight training, she competed with another student to become the first to gain enough flying hours to solo; she earned her commercial pilot's license shortly thereafter, in 1933.[27] Love was once a passenger in an airplane that "spun in," resulting in injury to the pilot and a cracked skull for her. Although she entered New York's Vassar College in the fall of 1931, she did not graduate, leaving school to pursue a career in aviation.

During the Great Depression, Love tried to get a flying job, but she was unsuccessful. Her parents sent her to New York City to secretarial school. While there, she flew as often as possible in nearby Newark, New Jersey. She eventually found work selling airplanes and flight lessons in Boston, simultaneously acquiring flying experience in many different types of aircraft. There she also met her husband, Robert Love, who had left college himself to develop his Inter-City Aviation business, now Boston's Logan Airport. Nancy Harkness and Robert Love were married in January 1936. During the early years of their marriage, Nancy Love worked in America's air marking program while at the same time testing the Hammond Y, a tricycle-gear airplane.[28] She also tested the Gwinn aircar. Although she considered the aircar a very safe model, she made her last flight as a Gwinn employee on June 14, 1938, flying the short distance from New York City to Danbury, Connecticut.[29]

In June 1940, Nancy Love helped ferry military airplanes. Landing in Maine, just short of the Canadian border, American aircraft were then towed across the line on land and, once they were in Canada, they were flown to embarkation points for delivery to France, which used them in patrolling before its surrender to Germany. Convinced that women like her could work in ferrying operations and thus alleviate the shortage of ferry pilots, Love wrote to Lt. Col. Robert Olds in May 1940, outlining her plan to gather highly experienced women pilots, their names culled from her assembled list of 105 women who held commercial ratings. In 1942 Love obtained permission to recruit women who could assist in moving military aircraft from manufacturers to ports for shipment overseas. Drawing on her list, she organized the first group of women in America to fly military aircraft, the Women's Auxiliary Ferrying Squadron (WAFS).[30] After her organization was merged with the training program begun by Jacqueline Cochran, Nancy Love became the commander of the ferrying division of the Women Air Force Service Pilots (WASP) until its disbandment in late 1944. In 1946, Love was awarded the Air Medal for her leadership of the WAFS.[31]

Nancy Love cherished and guarded her privacy, even while leading a public life; and she left almost no documentary trail, particularly concerning her experience leading the WAFS and the ferrying division of the WASP. She is best known through the commentaries of those who knew and worked with her, rather than from her own accounts, which are official rather than personal in nature. By examining documents in the possession of Love's family and interviewing her three daughters, Rickman has drafted the most complete biography

Nancy Love in her WAFS uniform at Newcastle AAB in Wilmington, Delaware. She chose its gray-green fabric and approved the design (The Woman's Collection, Texas Woman's University).

of Love currently in existence. Rickman notes that Love apparently felt pressured by the fact there was a group of women pilots Cochran was training in Houston. So she tried to hurry her own pilots toward transitioning to more advanced aircraft, and she paved the way for them by learning to fly all the latest aircraft models herself. One of her WAFS states that Love served as a model to gauge whether a woman of average size and strength would have any limitations in making the transition.[32]

By 1943, Love had made the first delivery of a pursuit airplane by a female pilot.[33] She believed that the behavior of the first women pilots must be absolutely impeccable, free of even the slightest hint of scandal. Love's administrative duties tended to prevent her from doing much ferrying work after her attempted trans–Atlantic flight with Betty Gillies. Concurrent with WASP disbandment in December 1944, Love was temporarily assigned to fly the Hump airlift by General C.R. Smith, the deputy commander of the Air Transport Command.[34] In January 1945, Love made a secret observation flight from Calcutta to China in a C-54, which delivered supplies to the Chinese and U.S. Marines. Technically, she was no longer eligible to fly at that time. The WASP had ceased operations on December 20 and General Arnold had ordered the women to stop flying military aircraft.[35]

Nancy Love was the mother of three daughters, the eldest born in 1947. The family moved to the island of Martha's Vineyard in Massachusetts in 1952. Love continued to fly on her own, often doing errands on the mainland in her single-engine, four-seater Beechcraft Bonanza. She was convinced that airplanes would become as common as the family car, and she preferred flying to driving on the mainland. Her daughters recall that their mother suffered from depression in her later years and developed a dependence on alcohol. Love succumbed to cancer on October 22, 1976, the WASP having named her "1976 Woman of the Year" in time for her to receive her award in the mail.[36] Among her effects found after her death was her list of women with commercial pilot's licenses in 1940 and news clippings about the women who had died under her command.[37]

10

The Pilots' Stories

Besides the life stories of the four women who played pivotal roles in ushering their peers into military aviation, the stories of the pilots they recruited are worth noting. This chapter presents the biographies of selected women pilots from Great Britain, the Soviet Union, the United States, and Germany, a country where a small number of women flew military aircraft independent of organized units. One of the women in this chapter was born in South Africa and another in Canada, but both flew with the British Air Transport Auxiliary. Alas, it is necessary to omit more names than can be included. That decision was subjective, but usually depended on how well the woman illustrates a particular aspect of the work the pilots performed. The entries are arranged by country according to the date it entered the war and then alphabetically. There is fairly abundant information on only one German woman, Hanna Reitsch, although there are six women known to have been pilots in Germany during the war.

Great Britain

Lettice Curtis (1916–)

At the age of seven, Lettice Curtis was sent to Benenden, a boarding school where she engaged in sports, especially tennis. She later attended Oxford, entering St. Hilda's School there in 1933.[1] Her autobiography does not mention the year of her birth. Curtis took her first flight as a young girl, a joyride she says left little impression on her. Her more lasting introduction to flying was the result of her post-university quest for employment. One day she was driving past Haldon airfield when a pilot, who had just landed, asked her for directions. While they chatted, Curtis inquired whether women could be pilots and he responded in the affirmative, setting her on a path to finding employment in aviation; in her opinion, that was far better than a desk job.

By June 1939, Curtis was noticing that civilian aircraft in England were being used in army cooperation work, which involved flying up and down a course so gunners could practice aiming their antiaircraft guns.[2]

In 1940, while employed by C.L. Air Surveys, she received a letter from Pauline Gower and decided to join the Air Transport Auxiliary, considering it a good way to contribute to the war effort. Curtis cites ATA's flying rules as follows: avoid interfering with active or passive defenses of the country; avoid risk of false air raid alarms; avoid risk of being shot down by their own defenses. She gives short shrift to Amy Johnson's disappearance in her autobiography, citing the resulting "flap" as the reason behind the disappointing cancellation of the one-year anniversary party celebrating the opening of the women's ferry pool at Hatfield.[3] Curtis flew so early in the existence of the ATA that pilot's notes had not yet been developed, so the ferry pilots gained information about airplanes any way they could. Sometimes this method led to problems, particularly if they were unfamiliar with an aircraft type.[4] In June 1940, the Royal Air Force handed over all ferrying duties to the ATA due to a shortage of fighter pilots. When the war situation worsened in autumn 1941, the women were given permission to fly all operational aircraft so that every ferry pilot could be employed to the utmost. The women entered the RAF's Central Flying School at Upavon for conversion training.[5] Curtis notes that she and her female colleagues were well aware that their performance was always under scrutiny and she says they tried not to bring limitations on the scope of other women pilots by their own actions.

Curtis writes that the ferry pools each had their own character, describing Ratcliffe thus: "In ATA generally, pilots gravitated to Ferry Pools which suited them both flying-wise and socially. No. 6 pool had collected together a band of relatively young, tough and self-assured pilots, many of them Americans, who had much more flying experience than us and liked to think of themselves as dead-end kids who could deliver their aircraft when even the birds were walking."[6] Curtis comments that she never felt at home posted at Ratcliffe and she even contemplated leaving the ATA altogether rather than continuing there.

Although she does not mention Jacqueline Cochran in an ATA context, Curtis does note that she stayed in Cochran's New York City apartment during a postwar visit arranged by Ann Wood and that she specifically remembers the "compass rose" design set into the aviator's hall floor.[7] She notes that the American women who arrived in Britain were posted to Luton, where they could learn about wartime flying conditions in the United Kingdom.[8] Her own daily life at No. 1 Ferry Pool, White Waltham, followed a typical plan. At 9:00 A.M., pilots arrived to pick up their ferry chits, laid out by the operations staff. If the chits weren't ready, the pilots might wait in the mess or a rest room. Expediency was the chief concern when making assignments to minimize the need for taxiing pilots back to their airfields. Taxi aircraft were flown by senior members of the pool or older and less active pilots of proven flying ability. Etiquette prevailed on the taxi aircraft, in which the more senior pilot was offered the controls. The operations staff sometimes designed the day's work based on their knowledge of the preferences of the pilots. Pilots had some latitude in the pace of their work, and Curtis preferred to fly first and eat later. The ATA medical officer managed to allocate a supply of 2 oz. Cadbury milk chocolate bars for the pilots each day, dispensed upon the presentation of a ferry chit.[9]

Curtis writes that some commanding officers would not allow women into their ferry pools, but others were more accepting, including Frankie Francis, the CO at White Waltham, who put her name through to train on the Halifax, a four-engine bomber.[10] Curtis relates one instance in which she landed a Halifax with the "assistance" of FIDO (Fog Investigation and Dispersal Operation), which consisted of a line of burners placed along the runway and designed to burn away the cloud or fog to a height of approximately 400 feet. As she circled over the airfield at Graveley, the operators turned on their burners. Curtis landed the Halifax, guided by the parallel lines of flame, her visibility not much aided by the smoke the fires produced, which had not yet burned long enough to clear the fog anyway. She comments that a lighter aircraft might have been affected by the turbulence created by the heat and rising air and might even have been pushed into the flames.[11]

The only accident Lettice Curtis experienced during the war was while landing a Typhoon that flipped onto its back, causing cuts to her head and leg. However, she writes, she was not so injured that she couldn't insist that she be taken to the ATA "home" hospital, the Royal Canadian Hospital at Taplow. One indication of the fatigue ferry pilots endured was Curtis' statement that she slept for most of the following week in the hospital, having flown virtually nonstop for four years.[12] During her recovery, she spent time with Margie Fairweather, one of the original eight Women's Section pilots, who was killed four months later due to engine failure while force-landing a Proctor.

In November 1945, after the ATA had ceased its operations, Curtis wrote to the airlines that were advertising for pilots. She was interviewed — out of curiosity, she thinks — but was not hired. She was offered the position of operations officer by the Ministry of Civil Aviation. Curtis wrote her history of the ATA, *The Forgotten Pilots*, which was published in 1971, although she states that she would gladly have left that honor to Pauline Gower herself, the founder of ATA's Women's Section.[13] Later in life, Curtis obtained her helicopter license; but, by 1995, she realized that she was not doing enough to maintain her pilot's licenses and decided to "call it a day."¹⁴ She published an autobiography in 2004 and contributed to Whittell's 2007 *Spitfire Women of World War II*. An RAF website provides an update on her current activities, noting that she received a commemorative badge for courage, as did fourteen other surviving female Spitfire pilots, in 2008.¹⁵

Helen Harrison (1909–1995)

Helen Harrison was born in Vancouver, British Columbia. Her family moved to England while she was in her teens and she attended boarding school in Britain, an attempt by her mother and father to exert some parental control over the lively teen. By age twenty-four, Harrison had married and was the mother of three children. She took her first flight in 1933 and found the experience thrilling. She earned her A and B licenses and in addition to flying the other pilots taught her to "drink, smoke, and swear." A newspaper story, written during her ATA service, cites Harrison joking that her presence among air force officers caused the men in Africa to learn Afrikaans, the Boer language, and the men in Canada to develop an interest in Iroquois or "Eskimo" so they could swear in front of her in a language she wouldn't understand.¹⁶ The reporter may have felt a need to highlight a woman pilot's feminine credentials, but that particular statement is hard to reconcile with the testimony of one of Harrison's former flight students, who claimed that he'd "never heard a lady use such colourful language."¹⁷

Harrison moved her own family from England to South Africa, and there she became the first woman in the British Empire to instruct on military aircraft.¹⁸ When Great Britain declared war on Germany, she returned to England and tried to join the RAF but was rejected. She went back to Canada early in 1940 and likewise attempted to join the Royal Canadian Air Force. She describes her frustration at being rejected a second time on the grounds of her sex, in spite of her proven ability and experience: "When I applied to the RCAF I was rejected because I wore a skirt. I was furious. I just couldn't believe it. I had 2600 hours, an instructor's rating, multi-engine and instrument endorsements, a seaplane rating, and the experience of flying civil and military aircraft in three countries. Instead they took men with 150 hours."¹⁹

Women in the RCAF could work as flight instructors, but not as pilots. Harrison assembled a women's flying class in Bob Redmayne's flying school in Canada.²⁰ She hoped her initial class of twenty-five women might form the nucleus of a Canadian Women's Auxiliary Air Force, similar to those in South Africa, England, and Australia. But after her plan for training women to ferry aircraft in Canada withered, Harrison instead seized the opportunity to join the women whom Jacqueline Cochran was preparing to fly with the Women's Section of the ATA, then gathering in Montreal.

A Canadian newspaper interviewed Harrison while she was on leave after spending sixteen months with the ATA. Harrison comments that the ATA women would have been glad for a chance to fly in combat, as the Russian women did, and although there was no chance that they would see that kind of action, she says they often discussed it. She stressed the need for "sound judgment" while flying in England due to a prohibition on instrument flying and the absence of radios in the new airplanes. Harrison trained on British Class IV aircraft and logged 500 hours. When she prepared to return to England after her leave of absence from ATA in March 1944, she was informed that she was no longer needed.[21]

After the war, Helen Harrison had difficulty finding aviation employment, but she did find work at the Dorval Airport in Montreal. One day she gave a ride to a wartime acquaintance, who was shocked to find a highly skilled pilot like her working on the ground as a driver. Harrison's friend helped her get a job demonstrating aircraft for the Percival Proctor Company in Canada. After that short-lived position ended, she moved with her three sons to Vancouver in 1947 and obtained a "float rating"; but the only possibility for employment there was as a flight instructor at various air clubs. When permanent managerial positions became available, the highly experienced Harrison was not considered for them.[22] According to Knapp's *New Wings for Women*, published in the immediate postwar period, Helen Harrison believed that the future for women pilots in Canada was brighter as a result of the war; but at the same time she acknowledged that each woman is ultimately responsible for her own success in aviation.[23] Render, who interviewed Harrison in 1983 and again in 1986, notes in *No Place for a Lady* that the otherwise voluble Canadian aviator had few words to say about her activities in the years between 1950 and 1969, the year she stopped flying.[24]

Amy Johnson (1903–1941)

Amy Johnson was a daring pilot whose determination, keen sense of direction, mechanical ability, and courage were celebrated beyond imagining in an era when few women drove cars and fewer still flew airplanes. Johnson was born into somewhat humble circumstances. Her family's business was started by her immigrant grandfather, who imported and exported fish in Hull. Born July 1, 1903, the year of the Wright brothers' historic flight, Amy Johnson's first taste of aviation was a joyride with her sister in 1926, an experience that failed to impress her. However, the adulation that Amelia Earhart received for simply being a *passenger* on a flight across the Atlantic may have caught Johnson's imagination, and it may have been one element in her decision to learn to fly. One of her biographers speculates that a second motivator was most likely her desire to impress her Swiss lover, who spurned her affections and eluded her dreams for marriage. Whatever her reasons, Johnson, who was working as a secretary, took her first flying lesson in 1928.[25] She joined the London Aeroplane Club at Stag Lane and there she became acquainted with Pauline Gower. Flying satisfied Johnson's appetite for thrills and adventure, and the flying club supplied the Moths she could not afford to buy. Flying lessons were expensive, easily costing her an average week's salary. Johnson soloed in 1929 after an unusually long period of dual instruction — over fifteen hours — considered close to the upper limit for normal student preparation.[26] Johnson's poor landing technique presented her a significant challenge. At a time when women were forbidden from entering the hangars where airplanes were serviced, she learned how to service the club's airplanes and found that she was as fascinated by the mechanics of flight as by actually flying. Her knowledge was largely self-acquired and often grudgingly dispensed by the male mechanics. However, Johnson, who liked to be known by her nickname "Johnnie," won the admiration of Jack Humphreys, a ground engineer. His decision to include her as an equal in the all-male context of the hangar gave her the apprenticeship she needed. It was Amy Johnson's ability to repair her own airplane that later made it possible for her to complete long solo flights across remote landscapes, even after her small airplane developed mechanical problems.

In 1929 Johnson decided to make aviation her career and she left her secretarial job. She studied engineering too and obtained her ground engineer's C license in December.[27] She later earned her ground engineer's A license, which complemented the one she already had, qualifying her to make preflight inspections of aircraft.[28] Johnson was already planning a solo flight to Australia when a reporter interviewed her for the *Evening News* in hopes of featuring her unusual presence in the hangar, a woman in a man's world. Although the article gave her valuable publicity, Johnson complained that the reporter had distorted the facts. He had crafted his story as others also would — by emphasizing his subject's glamour while giving short shrift to the gritty reality of her work. In preparation for the Australia flight, Johnson also honed her navigational skills, achieving yet another license. She needed the money to purchase an airplane and finance her flight, so she sought funding by courting sponsors. She solicited help from the director of Civil Aviation, Sir Sefton Brancker, who contributed to her venture along with her father, Will Johnson.[29]

In 1930, Johnson achieved overnight fame when she flew alone from

England to Australia in a green Gipsy Moth she'd named *Jason*.[30] Almost as soon as she started out, she was a news sensation and as such she was tracked by the press during her flight. By the time she landed in Darwin, she had achieved a celebrity status experienced by few. She received the Commander of the Order of the British Empire (CBE); songs and poems were written to commemorate her flight; and "Amy, Wonderful Amy" was suddenly a highly marketable commodity. Although her long distance solo flight was not a unique accomplishment — other pilots had flown the route by then — hers was the first flight from Britain to Australia by a woman. Consequently, Johnson's bravery was celebrated as much as her skill. She spent six weeks in Australia and was feted everywhere she went. While on tour, she met Jim Mollison, a Scottish pilot, whom she would marry in July 1932 and divorce in 1938. Johnson was sometimes perceived as grasping and aloof, but her biographers claim that she may instead have been out of her depth, unused to fame and often overwhelmed by the public's expectations. The *Daily Mail*, which had the exclusive rights to Amy Johnson's story, awarded her a large monetary prize; but they expected in turn that she would meet their requirements, including undertaking grueling publicity tours.

To maintain public interest and perpetuate her funding, Johnson tried always to exceed the adventure quotient of her most recent accomplishments. After her flight to Australia, she planned to fly to China via Siberia in an open-cockpit airplane during the winter, a plan no one else thought well advised and an excursion she began but did not finish, choosing to suspend the attempt in Moscow and continue it during the summer. She and Jack Humphreys flew across the Soviet Union to Tokyo in summer 1931.[31] A biographer speculates that Johnson's later flights may have been planned as much to escape the now relentless publicity as to maintain her focus on aviation and set new records.

Johnson, "the lone girl flyer," with Jim Mollison, "the flying Scot," set several flight records together, including flying across the Atlantic from east to west in 1933, aspiring to be the first husband and wife team to "conquer the Atlantic."[32] Because Mollison chose to fly their airplane, *Seafarer*, nonstop to their destination in New York rather than taking time to refuel in Boston, as Johnson suggested, they landed — out of fuel, fatigued and disoriented — upside down in a salt marsh beside an airfield in Bridgeport, Connecticut. After recovering from their injuries in a local hospital, the two were welcomed as the "flying sweethearts" in a New York City ticker tape parade.

As famous as she was for her aviation feats, Johnson was not a natural born pilot. But she had drive, naïve fearlessness, and the ability to perform mechanical work on airplanes, qualities that helped her set several aviation records. According to Babington Smith, Johnson was more interested in meeting new flying challenges than in perfecting her technique; and other pilots did not join in her public acclaim because none considered her especially competent. They joked that she "arrived" rather than "landed."[33] Often it was her innate sense of direction, combined with good fortune, that gave her the chance to tell the tale after a flight.

Johnson parlayed her love of adventure and willingness to take risks into instant fame and wealth, but maintaining her new way of life rapidly eroded that fortune. She considered joyriding as a potential source of income, but her letters show that she resented the lack of opportunities for females in aviation, the career that she loved. Great Britain established the Civil Air Guard on October 1, 1938. Although Johnson was interested in participating, she was not invited to apply. Her correspondence shows that she was miffed at being overlooked as a candidate for the CAG.[34] The organization was convinced that she would not be interested, based on the premise that Johnson had always sought payment for her work and CAG service was voluntary and unpaid. She would have liked a leadership position in the CAG, but lacking that, Johnson became a reporter for the *Daily Mail*, and in that role toured CAG training facilities. When the war began, Johnson was ferrying airplanes for the Portsmouth, Southsea, and Isle of Wight Aviation Company; but the hazards presented by the new defensive barrage balloons in the skies and the restrictions against females flying in wartime severely limited her opportunities to fly.

Johnson was not considered as a potential leader for the Women's Section of the Air Transport Auxiliary. However, she was one of the two pilots that Pauline Gower added to the

Amy Johnson in 1939. Her solo flight from England to Australia in 1930 made her an instant celebrity. Her aviation records brought her wealth, but they also exacted a price as she constantly sought new records to break. Johnson joined the ATA and was killed while flying an airplane over the Thames on a winter Sunday early in the war (courtesy Library of Congress).

original eight women in May 1940. At first reluctant to join the ATA, perhaps resentful about the selection of her old friend as its leader rather than her, Johnson was encouraged to join by Gower herself. She took the ATA flying test and was interviewed just like any other applicant, although she was so nervous at the prospect of competing equally with women to whom she was like a god that at first she feigned illness to avoid having to undergo a flight test. She submitted only after Gower assured her that such an examination was a mere formality. Johnson became a dependable and committed member of the ATA. Some of her colleagues initially thought that the famous aviator would be more liability than benefit to the organization. Johnson was perceived to be less amenable to training and more likely to expect special treatment than the average volunteer, and she attracted publicity wherever she went. At first, Johnson complained in her letters about the "girls' school" tone that she sensed at Hatfield, where she was stationed, and at the boredom of the work. When the "phoney war," immediately following England's declaration of war, changed into genuine (active) war, Amy Johnson appeared to change her opinion, and the work became more compelling. Her fellow pilots found her friendly, down to earth and dependable.

One cold and cloudy day in January 1941, Johnson set out from Squires Gate to Kidlington with an Airspeed Oxford, which was painted with a yellow belly to indicate a "friendly aircraft" to those below. Finding no opening in the cloud cover through which she could descend to a landing site, she parachuted from her airplane into the Thames River estuary. The *Haslemere*, a nearby ship whose crew had noticed her descent, attempted a rescue. Despite their efforts, including the eventual hypothermic death of *Haslemere*'s captain, who jumped into the frigid water to reach her, Johnson was swept under the ship's stern and the propeller. The accident spawned rumors about Johnson's final mission, as well as speculation about whether she carried a passenger with her, and if so, the mystery passenger's identity.

The Women's Section of ATA were about to celebrate their first anniversary, but the festivities were cancelled when news came that Johnson's airplane was missing. Johnson's initialed bags were recovered from the water, but neither her body nor her airplane was ever retrieved. Pauline Gower informed her friend's parents of her death and drafted an obituary. Gower, as a witness at the probate court hearing for Amy Johnson's will, testified: "In my opinion, she struck icing conditions and probably made attempts to get down to the ground, but was unable to break through the fog cloud. When her petrol had either been exhausted or she knew it was on the point of running out, she may have lost control of the machine because of the accumulation of ice, or decided to bale out as the only chance of saving her life."[35]

Dolores Theresa "Jackie" Sorour Moggridge (1922–2004)

In Pretoria, South Africa, young Jackie Moggridge, challenged her stepbrothers by announcing her intention to become a pilot. She consequently received the gift of a "joyride" for her fifteenth birthday, a flight that turned out to be a rough beginning in aviation for the frightened and airsick girl. Despite that, she requested another flight for her sixteenth birthday, more an attempt to save face than a desire to return to the air. Airsick once again, but also intrigued by this second flying experience, Moggridge began taking flight lessons over the objections of her entire family, who were convinced that the money ought to be spent on more acceptable feminine pursuits such as a finishing school, attending university or getting married. Moggridge describes her first solo: "In the most complete solitude I have ever experienced I joined the sky. Looking down at the earth receding into a blur of green and brown I sang and handled the air-craft carelessly. At last there was no instructor in front to comment acidly on the skidded turn or wayward airspeed. 'Look,' I cried. 'Everyone look up. It's me; Jackie Sorour. I'm flying, flying.'"[36]

At age seventeen, Moggridge earned her A license. Her next challenge was to parachute from an airplane, which she did, but the jump broke her ankle. In June 1938, Moggridge travelled to England on a quest for her B license, while her mother, more alert than her daughter to world events, unsuccessfully tried to convince her to come home if war broke out. In England, Moggridge added an assistant instructor's endorsement to her A license, and she found work teaching weekend students in a local flying club. One Sunday in September, Moggridge took to the air with a student. As she recounts in *Woman Pilot*, they took off "in peace at nine-thirty and landed in war at noon."[37] English skies emptied rapidly of all nonmilitary aircraft. From that time on, she deeply resented the fact that she'd been born female, her sex now an insurmountable impediment blocking her return to the sky. Told that the services of women pilots would not be required, she saw no option except joining the Women's Auxiliary Air Force, where she worked as an ambulance driver in the war's early months.[38]

Moggridge had too few flight hours at that point to join the ATA's Women's Section, so she trained for work in radar interception. When the war broadened and the need for pilots became acute, Moggridge reported to the Hatfield Aerodrome for an ATA interview and flight test, which was conducted by one of the eight original women pilots in a Tiger Moth airplane.[39] Several months elapsed and she began to feel hopeless about the possibility of a transfer because her work in the WAAF was considered too valuable to lose to reassignment to ferry aircraft. However, Moggridge finally received orders to report to the ATA and was interviewed by Pauline Gower. Reacting to Amy Johnson's notoriety, Moggridge rushed up to the startled celebrity to request an autograph at their first meeting, an action she later called her first gaffe.[40]

After the war, Moggridge joined the Royal Air Force Volunteer Reserve. She thinks that at first the RAF didn't know how to accommodate a female among the ranks of its pilots; but in time she was stationed at the Filton aerodrome near Bristol, the only woman among the fifty or sixty pilots there. Her training, despite the fact she had logged almost 3000 hours, was lacking in the skills needed to advance to full RAF Wings, therefore she only qualified for preliminary Wings. Moggridge also found work as a part-time pilot for a flying club, and she sometimes brought her young daughter along on her short joyriding flights. Moggridge says that she felt the effects of sexism throughout her life, because it often impacted her ability to pursue her love of flying, especially after the war ended.[41] During these years Moggridge noticed that former RAF pilots had easy access to civilian jobs in aviation. After a year in the RAFVR, she earned the commissioned rank of pilot officer, a status she considered more honorary than substantive but which she nevertheless cherished.

Moggridge soloed in a Meteor Jet aircraft and she would have liked to have been the British female pilot to break the faster-than-sound women's international speed record, competing against Jacqueline Auriol of France and Jacqueline Cochran of the United States. In 1953, in response to the petition she sent to Britain's secretary of state for Air, Moggridge was informed that such an endeavor "would serve no purpose."[42] A chance meeting with a former ATA pilot, who was a captain for BOAC, led to her employment by Air Services ferrying Spitfires to Burma. The connections she had made during her ATA service in that case landed her a job in the aviation industry during a time when women pilots were not especially welcome. In 1957 Moggridge became a pilot for Channel Airways. She was the mother of two daughters.[43]

Soviet Union

Valentina S. Grizodubova (1910–1993)

Valentina Grizodubova, who was born in 1910 in Kharkov (now Kharkiv, Ukraine), was the daughter of an inventor and aircraft designer, who in 1908 developed a primitive flying machine. The girl traveled with her father, learning geography along the way. Grizodubova graduated from the Penza Flying Club of the *Osoaviakhim* in 1929, and her acquaintance with Sergo

Ordzhonikidze, a friend of Joseph Stalin, gained her admittance to prestigious flying schools such as the Kharkov and Tula Advanced flying schools.⁴⁴ She joined the Maxim Gorky Escadrille, the first woman to be accepted, and flew throughout the country speaking to people in far-flung settlements and spreading educational pamphlets for the Soviet government. Stunt flying and parachute jumps were part of the airshow. Often it was their first contact with a woman pilot that sparked the desire to follow a similar course in hundreds of Russian girls.⁴⁵

Grizodubova was the pilot, along with navigator Raskova and copilot Osipenko of the women's crew that set a world (women's) distance record (6,450 kilometers) in the *Rodina* in 1938. Grizodubova's skill on that flight enabled the *Rodina* to belly land in the *taiga* after it ran out of fuel before reaching Konsomol'sk-na-Amure, its original destination. She received the Hero of the Soviet Union award along with her fellow aviators, the first Soviet women so honored, and she also received a prize of 25,000 rubles. She was elected a deputy of the Supreme Soviet, and was appointed chief of the Directorate of International Air Lines of the U.S.S.R. Civil Air Fleet. After the German attack of 1941, Grizodubova was elected president of the Women's Anti-Fascist Committee of Moscow.⁴⁶

During the Great Patriotic War, Grizodubova was the only female commander of an all-male regiment, the 101st Long-Range Air Regiment, consisting of about three hundred men who used passenger and mail-carrying aircraft to make bombing runs.⁴⁷ Later in the war, two women replaced their pilot husbands, who had been killed while serving with the regiment. Grizodubova managed to do more operational flying than any other regimental commander.⁴⁸

The men in her regiment respected their commander's skill as a pilot, but they appreciated even more her support of them as their leader, a risk for any commander to take during the Stalin era. Grizodubova even challenged Soviet authorities on behalf of her regiment.⁴⁹ In September 1942, the 101st Regiment was assigned to the Partisan Movement where it flew supply missions into the Bryansk Forest. Grizodubova then became a champion of the partisans, considering her most important task that of evacuating children from partisan detachments. Like Raskova, Grizodubova was a trained musician, and she accompanied the men's singing in her leisure moments, playing the piano. Recalled to Moscow in June 1944, Grizodubova was appointed deputy director of the Scientific Research Institute of Civil Aviation and chief of its Flight Testing Department.⁵⁰ According to Cottam, Grizodubova condemned Stalin's reign of terror and personally assisted five thousand people during that time.⁵¹

Lidya "Lilya (Lily)" Vladimirovna Litvyak (1921-1943)

Lidya Litvyak was born in Moscow on August 18, 1921. She was one of the young girls who were inspired by the exploits of famous women pilots like Grizodubova, Osipenko, and Raskova, the trio who set a record by flying nonstop across the Soviet Union in 1938. Photographs of Litvyak depict a delicate looking blond woman, an appearance in sharp contrast to her established reputation as one of the best fighter pilots of the war. During the Great Patriotic War, while serving in Stalingrad, Litvyak became the first woman to shoot down an enemy airplane.[52]

Litvyak has been termed gentle and shy, but she was also persistent and determined, qualities that helped her gain admittance to classes in an aero club at age fourteen, two years before she reached the age of eligibility. Litvyak trained at the Kherson Flying School, located in southern Ukraine on the Dnieper River outside of Moscow. At the time of the German invasion, she was an aero club instructor working near Moscow. Having been rejected by the field army, she heard that Marina Raskova was organizing a women's flight group and she volunteered. At Engels, Litvyak learned to fly the Yak fighter plane. She described her training in letters to her mother and young brother. Her father had been killed in Stalin's terror in 1937. Some speculate that her motivation to do well in the aviation regiments might have derived from her family's attempt to reclaim their previous status and to allay suspicion. Litvyak loved flowers, and her airplane's fuselage sported a white lily. She also developed the habit of planting little bouquets in the cockpit.[53]

Lily Litvyak became a lieutenant and a flight commander in the 586th Fighter Aviation Regiment. Later she was transferred with three other women to the 437th Fighter Regiment, which was stationed near Stalingrad, where she made her first two enemy kills. Litvyak was transferred again to the 9th Guards Fighter Regiment. An exceptionally skilled fighter pilot, she flew the Yak-1 airplane, a model she preferred. Finally, Litvyak and another woman, Ekaterina Budanova, were transferred to the 296th Fighter Regiment, which later became the 73rd Guards Stalingrad-Vienna Fighter Regiment. Litvyak continued flying the Yak, becoming a "free hunter," a skilled pilot allowed to seek out targets of opportunity. Litvyak's comrades describe her letters as a mix of mature patriotic sentiments and the naïve romanticism of youth.[54]

Myths surround Litvyak and her exploits and it is likely that some are inaccurate. The tale of her shooting down an ace in the German Luftwaffe, a holder of the Knight's Cross, has been revised by scholarship that proves that no one awarded the Knight's Cross was shot down on September 13, 1942, the day Litvyak became the first woman in history to bring down an enemy airplane. That day she scored two victories, and although she was an outstanding opponent, the story of the German pilot's shocked reaction to her description of their dogfight was undoubtedly embellished in subsequent retellings. Polunina's 2004 account of the 586th Fighter Aviation Regiment cites fewer personal and shared victories than is commonly recorded for Litvyak. Polunina, who was an aircraft mechanic for the regiment and is now the unit's historian and archivist, researched archival evidence, and she attributes a larger number of kills to Ekaterina "Katya" V. Budanova, Litvyak's friend and comrade.[55] Similarly, Lily Litvyak is often called the "White Rose of Stalingrad," although it was a lily she painted on the fuselage of her airplane, as the graphic representation of her nickname.[56]

Litvyak flew 168 missions and she is credited with twelve solo and at least three shared victories. She brought down an enemy observation balloon for which she won the Order of the Red Banner. She was well known along the Stalingrad front for her skill and courage in battle. Wounded once and hospitalized for several months, she recovered from her injuries and returned to her regiment to fight again. She was wounded a second time, but did not consider her injuries serious enough to prevent her from fighting. She took off for her fourth mission of the day on August 1, 1943, and failed to return. Searches for Litvyak's body were unsuccessful and rumors arose that she had run away. In 1979, however, schoolchildren found the remains of a small pilot that had been buried in a mass grave in the village of Dmitriyevka. The body was positively identified as Litvyak's, and the Soviet government, then convinced that she had been killed in action, awarded her the Gold Star and the title Hero of the Soviet Union on May 5, 1990.[57]

There is enough uncertainty about Litvyak's ultimate fate to raise some doubts. Polunina cites rumors claiming that Litvyak did not die in the wartime crash of her airplane, but was captured by the Germans. Some think that she managed to escape from the POW camp, where her presence was observed by Vladimir Lavrinenko, a fellow prisoner, and there is speculation that she availed herself of that opportunity to leave the Soviet Union.[58]

Anna Aleksandrovna Timofeyeva Yegorova (1918–)

Anna Yegorova was born in Volodovo, located northwest of Moscow, on September 23, 1918.[59] She was a member of the local Komsomol, and despite her mother's low opinion of the scheme, Yegorova learned to extol the virtues of collective farming. However she was expelled from the Komsomol for failing to convince *kulaks*, the "wealthy" peasants, of the benefits of collectivization.[60] When her membership was reinstated, it served as her ticket to work on the

Moscow subways, and in her later years, she spoke with great pride about the station she helped to construct. During their first days of working on the subway, Yegorova and the other girls were forbidden to enter the sealed pressurized underground caisson used by the workers because of the fear that subjecting them to pressurization would affect their ability to have babies. But Yegorova and her female coworkers protested and the ruling was rescinded.

Yegorova learned to fly in the Metrostroy Aeroclub, which required that she study gliding before taking lessons in powered flight. Yegorova continued to work on the metro, caulking tubing joints. On her first flight in an airplane, she took the controls, but worried constantly that she would fail, until she was reassured by her instructor that learning to fly takes time, just like constructing a subway. Yegorova soloed in a small open cockpit U-2. In her autobiography, *Red Sky, Black Death*, she compares today's commercial flight with the sensation of flying in the U-2:

> [Flying is] another matter altogether in a tiny training airplane like the U-2. You give yourself entirely over to the air — head, arms, shoulders. Dip your palm into the waves of air streaming by, and you can feel its harsh, icy thickness. Look around you. There's nobody near, as if you were alone in the universe. It's just the sky, you, and the airplane, compliant to your human will. It carries you higher and higher, toward the sun and stars, obeying your every command. You are its master.[61]

At the flying club, Yegorova also learned how to parachute from an airplane. She won the only female vacancy at the *Osoaviakhim* Flying School in Ulyanovsk, but was expelled on charges that her brother was an "enemy of the people." She was refused admittance when she reapplied. She joined another flight club and received a useful piece of advice from a Komsomol administrator who recommended that on future applications she should omit any mention of having brothers. Her competence in aviation gained her the only female spot in 1939 at the Kherson Flight School, in southern Ukraine on the Dnieper River.[62] After graduation, Yegorova became a flight instructor for an aeroclub in Kalinin (now Tver) northwest of Moscow.

On one of her days off, a Sunday in June, while picnicking in the countryside with friends, Yegorova heard shouts announcing the arrival of war. Impatient with inactivity six weeks later, and convinced that her aeroclub would soon need to retreat, Yegorova took the train to Moscow. She asked the local Osoaviakhim colonel to send her to the front, but he was already inundated with similar requests, and chose instead to assign her to a different aeroclub. Yegorova journeyed there by train but found the club had been abandoned. Waiting for the arrival of the personnel chief so she could ask him for another assignment, she met an officer who needed pilots. Because of that meeting, Yegorova joined the 130th Air Liaison Squadron of the Southern Front.[63] Her role was to fly a tiny U-2 airplane to deliver supplies. Yegorova cites several nicknames for the plane, including the pilots' own "little duck."[64]

A tale of what she terms "a routine day" constitutes her chapter titled "Abandoned." She was to transport a liaison officer carrying operational orders for soldiers in a little village. When they landed, the village was under attack. The officer told Yegorova to wait for him near the airplane while he investigated. Yegorova waited hours as German fire came ever closer, finally rocking her tiny plane. Determined to save the airplane, she commandeered a fleeing truck driver by shooting out his tires and commanded him to spin the propeller so she could take off. Everyone in her squadron had given up on her return, except for her mechanic, who set a small fire on the landing strip to guide her home. The officer she had transported claimed that he left her because he had to carry out his delivery of retreat orders to the (already fleeing) soldiers. He begged Yegorova's forgiveness when he heard she'd managed to escape. She writes of it: "*What kind of officer, and what kind of gentleman*, I thought, *would leave a woman to die like that?*"[65]

Letters from her mother caught up with Yegorova months after they were written. Under those conditions, families never knew whether their loved ones were still alive. Some of her missions involved flying in winter whiteout conditions to bring tactical information to officers when radio communication had failed. Yegorova's little plywood airplane was once set afire in an air attack. She managed to land and hide from the pursuing fighter, but she was badly burned. She lost her airplane but returned to her squadron. Yegorova writes that she never had a problem with the men. If she was courted by potential suitors, she was always careful to converse with them only in public, thus conveying a message of being disinterested in love. She describes her situation: "It was difficult being the only woman among so many men. Sometimes I just wanted to have a heart-to-heart talk with someone. But I had to hold those feelings inside."[66]

In early 1943, Yegorova was transferred to the 805th Ground Attack Aviation Regiment where she had female navigators and ground crew. One source claims that Marina Raskova repeatedly asked that Anna Yegorova be reassigned to her own flight group, but those requests were refused by Yegorova's commanders, a fact unknown to the pilot during the war.[67] Yegorova does not mention Raskova in her autobiography and remained the only woman pilot in her regiment. She flew the one-seater *Shturmovik*, or Il-2, an airplane she and other pilots considered beautifully designed in the battle for fighting the Taman Peninsula in formation with other airplanes, making several sorties every day to attack German front lines. She also flew a double cockpit Il-2 carrying a tail gunner, who was sometimes female. Yegorova thought the tail gunners' job was truly terrifying because they sat with their backs to the pilot, shooting without cover at airplanes that were shooting at them while their own airplane was constantly tossing about.[68]

Yegorova's regiment steadily worked its way east into Poland. She was shot down and captured in August 1944 and spent five months in a German concentration camp there. Liberated by the Russians in early 1945, Yegorova was accused of treason because she had been captured. She was interrogated by Soviet government and stripped of her awards. There was only one hope for Soviets who had been captured, "rehabilitation." Yegorova fought to regain her good name for many years after her wartime imprisonment. Eventually she was reinstated as a Party member and, thanks to her combat friends who sought out the records of her awards during the war, received a Gold Star Medal (HSU) and the Order of Lenin in May 1965. Yegorova also credits her friends with making it possible for her to obtain an apartment in Moscow.[69] At the end of the Great Patriotic War, Yegorova married V.A. Timofeyev, the commander of the Soviet 197th Attack Air Force Division, and became the mother of two sons.

United States

Ann Baumgartner Carl (1918–2008)

Ann Carl spent her girlhood in Plainfield, New Jersey. One day the famous aviator, Amelia Earhart, visited Carl's elementary school and stirred the young student's imagination with descriptions of the adventure of crossing the Atlantic Ocean by air in 1932. Carl's maternal family came from Yorkshire, England. She and her mother were visiting relatives there in 1939, just before the country declared war on Germany. The two women obtained passage back to America on a Dutch steamer on September 2.[70] Her British family ties gave Carl a strong sense of connection to England's plight during the Second World War. Although she wanted to learn how to fly, she just missed the opportunity that had been open to females in the CPTP, so she had her first flight lesson in a yellow Piper Cub at an airfield in Basking Ridge, New Jersey.[71] In 1940, Carl soloed after only eight hours of instruction. She arranged to get more flying time by sharing the purchase of a Piper Cub with another pilot because she wanted to accumulate the 200 hours necessary to obtain a commercial license, a step in making aviation her career. Carl joined the Civil Air Patrol, where she did what she terms "serious flying."[72]

While employed as a *New York Times* reporter, Carl became aware of Jacqueline Cochran's WASP program. She applied and was among several young women Cochran interviewed at her New York City apartment. Carl remembers that the process included completion of a questionnaire from Cochran's assistant, Ethel Sheehy. Carl was accepted into class 43-W-3 and she reported to Aviation Enterprises Ltd. at Houston Municipal Airport; but an illness prevented her from graduating with her class, so she finished with class 43-W-5. Carl and her classmate, Betty Greene, were assigned to the tow target squadron at Camp Davis in North Carolina, replacements for two WASPs who had been killed in accidents there. Carl notes that she and Greene occupied the dead women's quarters.[73] Later, both women were assigned to temporary duty at Wright Field Air Base in Dayton, Ohio, where they were charged with designing and testing a relief tube for females, similar to the one for males, a concept more feasible in theory than in reality, according to Carl, given the different construction of men's and women's undergarments. Carl also tested clothing designed for high-altitude use. When that temporary assignment ended, she asked for a permanent transfer from Camp Davis to Wright Field, a request that was approved.

At Wright, Carl was assigned to the Fighter Flight Test Branch and also to the Bomber Flight Test Branch. She became a test pilot for experimental American and foreign aircraft, such as the British Lancaster and Mosquito bombers, and the German Ju-88. She conducted refueling tests and did off-center flying in an XP-82 "twin Mustang." She conducted weight and distance tests in a B-29, preparation for the atom bomb mission over Japan. While Carl worked as a test pilot for the WASP in Ohio, she was frequently asked to accompany the airfield's well known namesake, Orville Wright, at various social occasions during his visits to Wright Field. The two pilots could easily talk "aviation." Wright encouraged Carl to consider flying a jet aircraft. In her autobiography, *A WASP Among Eagles: A Woman Military Test Pilot in World War II*, Carl notes that General Henry H. "Hap" Arnold had learned to fly with the Wright brothers.[74] Carl made an evaluation flight in the first jet plane with a pressurized cockpit, the RP-47E, and for a period of ten years she was the only woman who had flown a jet, the YP-59A.[75]

After WASP disbandment, Carl remained at Wright hoping to work in a new capacity, using her writing skills to tell the stories of wartime fighter pilots who were testing new aircraft there. The time was not propitious for that kind of endeavor, so she left the Wright base in February of 1945. Carl married and became the mother of two children. She and her husband pursued long-distance sailing, and she took on the role of navigator. When her children were school-aged, Ann Carl returned to aviation as a flight instructor.

Teresa D. James (1914–2008)

Teresa James grew up in Pittsburgh, Pennsylvania. She took her first solo flight at the Wilkinsburg Airport, the improbable outcome of her effort to overcome a strong fear of flying that she had developed after witnessing two serious aviation accidents.[76] One of the mishaps involved her brother, who crashed his airplane in 1931 and afterwards spent several weeks in the hospital. James would become an excellent but always extremely careful pilot, and she liked to compete with her counterparts to see who could best "grease" a landing. She first rode in an airplane as a passenger when a handsome pilot named Bill Angel invited her to take a ride; although she had agreed to the flight, she recalls that she found the experience terrifying. Angel moved on to another airport, his absence pushing the nineteen-year-old James to take flight lessons because she was convinced that flying by herself to his new location would be an excellent way to get his attention. James later described her feelings about being airborne: "The fear never really left me to the point where I said I can be sure I can go up there and get back down. I was always on top of flying; I was fright conscious. Some people I talked to were never afraid. Maybe I was an oddball, but I was absolutely terrified."[77]

James gained enough confidence to earn her private pilot's license in 1934. She worked as a stunt pilot to earn enough money to purchase additional flying time.[78] James learned to put her small airplane down practically anywhere so she performed her stunts without a parachute. She was sworn in as Wilkinsburg's first female airmail pilot in 1938, and earned her commercial transport license in 1941.[79] She also worked as a flight instructor, occasionally giving lessons to the man who would become her husband, George "Dink" Martin, who himself became a flight instructor and a bomber pilot in Europe during the war. James organized a unit of the Civil Air Patrol at Pennsylvania's Johnston Airport. Having logged 2,254 flight hours by then, she caught the attention of Nancy Love, who sent her a telegram inviting her to join the new group of women ferry pilots.[80] James accepted and traveled from Pittsburgh to Wilmington, Delaware, by train, sharing a cab to a local hotel with two other prospective WAFS, Helen Mary Clark and Aline Rhonie.[81]

In *On Wings to War*, James describes the difficulties that pilots encountered when they flew toward the West Coast of the United States. Coastal airports were camouflaged to deter enemy reconnaissance. Both weather forecasting and radio navigation were fairly primitive, and although both improved during the war years, navigation was still a challenge. Interviewed for the film *Women Combat Pilots*, James says that "jaws dropped" whenever she exited a plane at every base she flew to during the war.[82] James was the

WAFS Teresa D. James wearing flight gear and a parachute. Photogenic and personable, James was featured as an exemplar of the American woman pilot in wartime (The Woman's Collection, Texas Woman's University).

first WAFS pilot to fly a PT-19 from coast to coast, in 1943.[83] She ferried the airplane from the Fairchild factory in Hagarstown, Maryland, to Hollywood, California, for the famous aviator Paul Mantz to use in *Ladies Courageous,* a film some sources claim James herself inspired. A natural extrovert, James enjoyed the celebrity accorded her at Burbank Field, where she met many Hollywood stars of the era, including Bob Hope and Marlene Dietrich. James also appeared on the February 6, 1943, cover of *Look* magazine.[84]

James was one of the women Cochran selected for Army Air Forces Tactical School in Orlando, Florida, in 1944, while WASP militarization was still a possibility.⁸⁵ As a WASP, James was stationed at Farmingdale, New York, on Long Island and she was the pilot next in line to ferry the airplane that became known as "Ten Grand," the 10,000th P-47 produced at the Republic aircraft factory, on September 20, 1944.⁸⁶ After ferrying the fighter to Newark, New Jersey, James slipped a note into the plane's flight report wishing good luck to its first combat pilot. Lieutenant Arthur E. Halfpapp wrote a letter back, describing that particular airplane's performance and, by the way, noting several engine defects! For her flight that day, James was awarded a lifetime membership in the P-47 Thunderbolt Pilots Association.⁸⁷

Married in August 1942, James received news in June 1944 that her husband was "missing in action" over France. Decades passed before she learned the details of her husband's fate from local French eyewitnesses to the crash of his airplane.⁸⁸ After WASP disbandment, James offered her services to the Chinese Air Force, but her request was turned down.⁸⁹ She also approached American domestic airlines, but they would not hire women pilots. She decided to work as a flight instructor and did some air racing. She also managed her parents' flower shop. Teresa James was commissioned a major in the Air Force Active Reserve in 1950, and she retired in 1976.⁹⁰ She was a strong advocate for WASP militarization in the 1970s, and donated her WAFS uniform for display by the National Air and Space Museum of the Smithsonian Institution.⁹¹

Hazel Ying Lee (1912–1944)

Born in Portland, Oregon, Hazel Lee held dual Chinese and American citizenship. According to her friends and family, Lee was the quintessential example of a modern Chinese woman of the interwar period. Nevertheless, she experienced incidents of prejudice against her Chinese heritage, reinforced officially by the Chinese Exclusion Acts of the 1930s. Chinese women were not allowed to work in professions with high visibility — in offices, for instance — so Lee found employment as an elevator operator.⁹²

In 1931, Lee took her first flight in a friend's airplane and loved the experience. She earned her private pilot's license in October 1932, one of the first Chinese-American women to do so; and she joined the Portland (Oregon) Flying Club, one of only two women among the club's thirty-two members. When Japan launched an attack on China, Lee responded to the Chinese appeal for help in America by volunteering for the Chinese Air Force, but she was rejected because the Chinese military thought females would be "erratic in combat."⁹³ Disappointed by that rejection, Lee found work as a commercial pilot while she lived in Canton with her mother and sister, staying until the city was bombed by the Japanese in 1937. Leaving their possessions behind, Lee, along with her family and her friend Elsie Chang, walked to the railroad station and took the train to Hong Kong, where they all became war refugees. However, Lee grew restless with that forced idleness, so she returned to America in 1938 and helped send armaments to China from her home in New York.[94]

Lee entered WASP class 43-W-4 and her younger brother, Victor, joined the U.S. Army. While she was training in Sweetwater in 1943, Lee married Major Clifford (Lin Cheung) Louie, a pilot in the Chinese Air Force who had been wounded in China and was recovering from his injuries in the United States.[95] After her graduation, Lee was stationed at the Romulus Air Base in Michigan. One of her favorite airplanes was the new P-51 fighter. Lee attended officers training school in Orlando, Florida, along with several other WASPs in 1944. Her WASP acquaintances report that Lee was fun-loving and popular. One of her fellow WASPs drew cartoons depicting her in moments of relaxation, smoking, drinking, eating, and having fun with her fellow fliers. She was a good cook who guided her WASP friends to Chinese restaurants and assisted them with their orders. Lee painted her friends' nicknames in Chinese characters on their airplanes. In the film *A Brief Flight*, Lee's sister remembers that before Lee's death, she seemed "down" about the long duration of the war.[96] Lee embarked on her last delivery of a military airplane on November 23, 1944, flying from a refueling stop in Fargo, North Dakota, to the airbase at Great Falls, Montana. WASPs delivered the Lend-Lease P-38s intended for the Soviets and Lee was one of many pilots approaching the runway that day. As she descended, a second P-38, just above her, descended from a different direction. The tower ordered both pilots to pull up and go around again, and although Lee followed that order, the other pilot did not. The two airplanes collided and exploded. Although Lee was pulled from her burning airplane, the heavy winter flying suit she wore kept smoldering inside, spreading her burns. Lee was transported to the base hospital, where she died two days later. On November 25, 1944, Hazel Ying Lee became the last of the thirty-eight WASPs to die in service to her country.[97]

Three days after Hazel's death, Victor Lee was killed in combat in France. Their sister recalls that the family chose a burial spot overlooking the Columbia River in Portland, which was at first denied to them because they were Asian.[98] Florence, their older sister, fought for the right of her siblings to be buried in the Riverside Cemetery, and she threatened to take her plea to the United States War Department because her brother and sister had both died while serving their country.[99]

WASPs Hazel Ying Lee (left) and Autumn Geneva Slack at Avenger Field in Sweetwater, Texas. Lee was the last of the thirty-eight women pilots killed in the United States while serving as WASPs (The Woman's Collection, Texas Woman's University).

WASP Pat Dickerson, who had accompanied Lee's body to Portland, also retrieved and drove her automobile from Lee's home base in Romulus to her relatives in Forest Hills, New York.[100] WASP Fran Snyder Tomassy was the pilot who had landed just ahead of Lee in Great Falls and WASP Kay Gott observed Lee's collision from her own airplane, which was still in the air.[101] WASP Sylvia Dalmes Johnson packed up Lee's belongings after her death.[102] Gott was deeply troubled by Lee's accident and conducted a documentary investigation of her own, resulting in her book, *Hazel Ah Ying Lee*.

Helen Richey (1909–1947)

Helen Richey was the first female commercial and airmail pilot in the United States. Richey grew up in McKeesport, Pennsylvania, and as a young girl she was accepted onto the boys' sandlot baseball teams because she was a formidable opponent who could hold her own with her male contemporaries.[103] Richey learned to fly during America's interwar barnstorming years. She was the copilot with Frances Harrell Marsalis in a record-setting, nonstop flight which took place entirely in the skies over Miami, Florida, in December 1933. Richey and Marsalis flew in circles for ten days in hopes of breaking the women's endurance record, held by Marsalis and another pilot. The women's endurance record at the time was eight days, while the men's record stood at twenty-seven days. The women received food and fuel from a supply plane, which made a total of eighty-three mid-air contacts. The stunt was funded by the Outdoor Girl cosmetics company. Officially, Richey and Marsalis were aloft in their airplane, the *Outdoor Girl*, for 237 hours and 42 minutes, and they covered 23,700 miles in their monotonous circuit. Their fabric-covered airplane consumed eight tons of gasoline.[104]

In a bid to outshine a competitor, James D. Condon, the president of Pennsylvania's Central Airlines, hired Helen Richey when she applied for a copilot's position with his airline in December of 1934. Condon made sure that he reassured the United States Department of Commerce that he was hiring a woman only for publicity purposes and did not plan to retain her (in a pilot's capacity) for long. Richey's first flight on January 1, 1935, was touted as a breakthrough for women in aviation. Female pilots were a rarity in the early days of aviation and Richey seemed perfect for the job of bringing attention to Central Airlines, having earned notoriety for her recent Miami stunt.[105] The job would be to carry both passengers and the mail. The fact that a commercial airline had hired a woman pilot hit newspaper headlines across the country and made Richey an object of great curiosity, although she was not allowed to perform the takeoff during her first flight. The airline

Helen Richey sits in a cockpit in 1932 before an air race. Richey was the first woman pilot for a commercial airline in the United States, and she served in both the British ATA and the American WASP. After the war, she found scant opportunity to work as a pilot (courtesy Library of Congress).

would not let Richey fly in bad weather either, because that was considered too physically demanding for the female physique. Still, one newspaper headline crowed, "Helen Richey Conquers Last Stronghold of Men in Aviation, Gets Job in Cockpit."[106]

Although she was working as an airline pilot, Richey was refused membership in the Airline Pilots Association, and furthermore, the company's other pilots protested her hiring to the United States Commerce Department. They claimed that the idea of a woman being a pilot was preposterous because women lacked the strength necessary to handle an airplane, especially in bad weather; moreover, with this example, more women might be encouraged to learn to fly, and then they would take jobs away from men! The level of publicity surrounding Richey's hiring made it practically impossible for Central Airlines to fire her outright, so the company reduced her schedule instead. Realizing that she was underemployed when compared with her male colleagues, the first female commercial pilot resigned her position after eight months.

Richey was acquainted with Amelia Earhart and several other notable female pilots. Earhart launched a protest against Richey's treatment by Central Airlines, and she railed against the refusal of the pilots' union to accept Richey based on her sex alone. The pilots' association countered Earhart's criticism by stating that they were required to follow their bylaws, which restricted the group to men only. They pointed out that they had asked Richey to apply for membership a second time, an offer that she had refused.[107]

Richey turned down subsequent offers from other airlines, and she joined the air marking initiative of the Bureau of Air Commerce in December 1935.[108] Her job was to convince local officials in New England to allow the name of their towns to be painted in bright orange on preselected rooftops, or, if there was no nearby airport, to substitute the name and distance to the nearest landing possibility, as an aid to aircraft navigation. Eventually, Richey was put in charge of air marking for the northwest United States and her base of operations switched to San Francisco.[109]

Richey worked with the Bureau of Air Commerce until 1937. During that period, she also attempted new aviation records. One of them was the women's altitude record (18,000 feet) for airplanes weighing less than 440 pounds.[110] In 1936, Richey and Amelia Earhart flew as a team in the Bendix Race and finished fifth.[111] It is likely that while she worked in the air marking program Richey became acquainted with Nancy Love, who was working in the program.[112]

In early 1940, Richey trained civilian pilots under the authority of the Civil Aeronautics Administration, one of the first women to do so. She also trained other flight instructors, racking up a total of 1500 hours by June 10, 1941.[113] Jacqueline Cochran telegrammed Richey in January 1942, offering her a spot in the Women's Section of the Air Transport Auxiliary in England. Richey arrived in Liverpool with a few other American women on March 24, 1942.[114] Cochran asked her to manage the American women still serving in ATA when she returned to the United States to set up her own women's program. Some sources claim that the ATA, which had no tolerance for carelessness, sent Richey home for damaging too many airplanes, and she did crash land a Spitfire on one occasion.[115] However, her biographer maintains that she resigned her ATA position early in 1943 so she could return to America and assist her seriously ill mother. A *McKeesport Daily News* article quotes Richey describing the thrill of Mrs. Roosevelt's visit to the Women's Section while she was still working there.[116]

Accepting Cochran's standing offer to the ATA pilots from America, Richey joined the WASP upon her return, and, despite her considerable flying experience, she underwent the training program required for all WASPs, graduating on September 11, 1943.[117] Richey was a member of class 43-W-5, the first one to arrive and train entirely at Avenger Field in Sweetwater, Texas.[118] Richey's classmates included Ann Carl, Yvonne Pateman, and Marian Stegeman. She was assigned ferrying work at Newcastle County Airport in Wilmington, Delaware, and she was transferred to Fairfax Field in Kansas City, Kansas, where she trained on heavy bombers and cargo planes.[119]

Helen Richey never married, though she was engaged briefly to a friend and fellow McKeesporter, Jack Soles, in 1939.[120] Her niece, Amy Gamble Lannan, claims that Richey "had a phobia about marriage, children, and the little vine-covered cottage in suburbia. Right up to the day of my wedding, she kept asking if I were sure I was doing the right thing. She always felt there was more to life for a woman than being a wife and mother."[121]

When the WASP disbanded, Richey occasionally ferried surplus wartime aircraft, but she was never again fully employed. She moved to New York City in the summer of 1945, and it was there, on January 7, 1947, that she was found dead from an overdose of sleeping pills in her sparely furnished and locked apartment by her former WASP classmate Mary Parker.[122] Richey's suicide at age thirty-seven culminated a period during which she showed signs of severe depression. One source quotes her as claiming, "When a girl reaches 37, her flying days are over."[123]

The Pennsylvania State House of Representatives, on June 11, 1947, resolved to honor Richey as "an aviation pioneer and one of the state's outstanding citizens." In 1954, McKeesport named a baseball diamond, in its Renziehausen Park, Helen Richey Memorial Field. Although the field's bronze plaque displays an image in relief of Richey's uniform-clad portrait flanked by two airplanes, which was clearly copied from her official ATA photograph, its dedication acknowledges neither her service for the British ATA nor for the American WASP.[124]

Adela Riek Scharr (1907–1998)

Adela Scharr was the daughter of a policeman in St. Louis, Missouri. Scharr became a schoolteacher, and managed to save the sum of $2500, money she decided was more sensibly invested in a two-family brick flat than in the purchase of a second-hand airplane. Scharr took flying lessons in half-hour increments, over a period of five years, to amass the hours necessary for a commercial license. She was married by then and married women were not allowed to teach in midwestern schools during the 1930s. So she found work as a flight instructor in the Civilian Pilot Training Program at St. Louis University. Scharr's name attracted the attention of Jacqueline Cochran, who contacted her about an opportunity to join the women going to England to fly with the ATA, an offer Scharr declined. Six months later, when Nancy Love sent another telegram, America was at war, Scharr's husband was about to ship out with the navy, and she felt that the time was right for her to serve too, so she accepted.[125]

Sisters in the Sky is the title of a two-volume account of her experience in the WAFS and later in the WASP. The level of detail Scharr provides was possible only because she had asked her husband to keep all the letters she sent him during the war years. Augmenting those letters with her personal notes and flight logs, she crafted a practically day-by-day account. Scharr recalls an early conversation with Pat Rhonie, one of the first WAFS who had also been in the ATA, as the two women waited to be flight tested in Wilmington. Scharr describes how she found her own way to the airfield and notes the financial hardship she faced when she had to hire a taxi and stay in a hotel after the weather closed in and scrubbed her scheduled flight test: "Lt Tracy pointed me toward a telephone so I could call a cab. As I walked down the hall, a wave of utter loneliness engulfed me. The perfunctory and cool politeness of the few people I'd met stung me. How could the Ferry Command wire me to come, to leave all that was dear and familiar to me, and then not even care if I had a place to lay my head? Was the angelic-looking Nancy Love really thoughtless? Well, I'd take that flight test and I'd see."[126]

Never quite losing the habits she'd acquired as a schoolteacher, the new member of the WAFS thoroughly read specification sheets for every aircraft she might fly, and she never rushed through her ground checks. Scharr attributed her success in aviation to her painstaking study habits. She used a mnemonic to remind her of the checklist of tasks in the Pilot's Information File: **B**ut **T**hat **A**ll **G**ood **P**ilots **M**ust **L**and **F**inally, jogging her memory about

WAFS Adela Riek Scharr at Newcastle AAB in Wilmington, Delaware. After the war, Scharr wrote a detailed history of the WAFS and WASP based on the notes and letters she wrote during her service (The Woman's Collection, Texas Woman's University).

Boost (fuel for power), Trim (the balance), Altimeter, and so on, before every takeoff.[127]

Scharr's modest family background was very different, especially in her own mind, from the wealthier backgrounds of her WAFS cohorts. Most of the women selected by Love came from well-heeled families because flying was most commonly a hobby for the well-to-do. Women like Scharr, who had fewer resources but were determined to get into the air, could usually find a way to fund their flying lessons, so she was by no means unique. But among the group of WAFS, Scharr seemed keenly aware of their differences.

Love chose Scharr to assume command of the WAFS stationed at the Romulus Army Air Base.[128] The experimental nature of the program is aptly illustrated by Scharr's recollection of the quasi-military manner in which command assignments were made. Scharr's highly detailed account of the women's personalities and their apparent motives often casts an unflattering light upon her fellows. Scharr seems to have felt manipulated by Nancy Love as well as by the other WAFS.

Scharr developed a serious illness that kept her hospitalized during a significant portion of her time in service. When the WASPs disbanded, she was still recovering from that illness, but afterwards she transported surplus training planes within the United States. When the war ended, she contacted the head of the Monsanto Chemical Corporation of St. Louis with a plan for converting surplus military airplanes into executive aircraft for transporting people or materials. Her idea was well received, but not if women pilots would be flying the airplanes. A Monsanto executive claimed that the men's wives would never stand for their husbands being accompanied by a woman on their trips.[129] Scharr resumed her flight instructing. In 1948 women became eligible for service in the Reserve, and in the newly independent United States Air Force. Former WASPs qualified to be commissioned officers in the Reserve.[130] Scharr joined the Air Force Reserve in June 1949, at the rank of major. The law forbidding married women to teach full time had changed by then, so she returned to teaching. In 1961, at the age of fifty-four, Scharr got an opportunity to fly a four-engine C-135 jet, proving the capacity of a woman to fly it. She submitted her report on the experience to General Joe Kelly, Military Air Transport Service (MATS) commander, who was curious about whether a woman could fly the C-135:

> There was no problem for either sex when conditions were normal; even a brainy twelve-year-old with sensitivity could fly it. As with men, I'd say that she should have good musculature for the emergencies that might arise and therefore be taller than average. I thought that a woman sixty-eight inches tall and weighing at least 140 pounds should be adequate in any situation. Her

intelligence must be high, her reflexes and mental responses quick, her attitude alert, and she must be willing to over-learn all the systems and details involved with flight. She had to keep poised and able to think out problems even before she was faced with them instead of leaving it to luck or chance. These characteristics are an individual matter; they are not masculine prerogatives.

She added a postscript: "I should have added that the men in power must be agreeable to allowing such a woman to have her chance."[131]

Adela Scharr retired from the Air Force Reserve in 1967.[132] She was active in the WASP effort to gain veteran status in the 1970s.

Evelyn Sharp (1919–1944)

It was not until she was nearly grown that Evelyn Sharp learned that the woman she'd always called "Aunt Elsie" was really her mother, divorced from Evelyn's young father after a very brief marriage, and a woman who had done her best to provide for her infant. In December 1919, a childless couple legally adopted Sharp, changing all but her middle name but continuing to maintain their child's connection with her "aunt." During the years of the Great Depression in Ord, Nebraska, a man named Jack Jefford opened a flying school, but he had fallen behind in his room rent payments to Sharp's family. In exchange for his rent payment, Jefford offered to provide flight lessons for their daughter. This was how Sharp took her first flight in February 1935. She soloed after thirteen months.[133]

Sharp was a "natural pilot" and she soon became a minor celebrity in the state of Nebraska, an attractive young woman with real skill in a field that was dominated by men. Sharp joined a local chapter of the Ninety-Nines in 1937, one of only 444 licensed female pilots in the country.[134] She wanted a commercial license but found it difficult to acquire the necessary 140 hours of flight time. Although her family could not afford to purchase an airplane, the businessmen in Ord bought Sharp a new Taylor cub, their motivation partly to help a deserving young person and partly to promote Ord and support its North Loup River diversion project.[135] True to her end of the bargain, Sharp took part in a beauty pageant, judged a model airplane building contest, and distributed promotional postcards for the dedication of the new Arrasmith Field in Grand Island.[136]

Sharp failed the written portion of her first transport exam because she needed additional ground instruction. Her further education in this area was financed by the townspeople of Ord, who held a benefit dance so Sharp could attend the Lincoln (Nebraska) School of Aviation.[137] Sharp lost her first airplane when her father could no longer make his portion of the payments. She continued to make local public appearances, and finally obtained her commercial license at age eighteen. She purchased a used Curtiss Robin and gave airplane rides to help pay for it. On her airplane she painted the following words: "Miss EVELYN SHARP, Nation's Youngest Aviatrix Government Com. License 34711." Her barnstorming work at one time served to support Sharp, her parents, and one employee.[138]

Evelyn Sharp was determined to earn a living through flying, but considered barnstorming an unreliable source of income because it was so dependent on the good weather of summer. She applied to the Civilian Pilot Training Program and signed a contract to train fifteen students in the summer of 1940. She worked long hours instructing, sometimes taking no days off. She was in California working for the CPTP on December 7, 1941. Her flight school program then moved to Lone Pine, California.[139] There she received a telegram from Jacqueline Cochran inviting licensed female pilots to fly with the ATA. By that time, rumors were already circulating that a similar unit would soon be formed in America, so Sharp decided to wait. On September 5, 1942, she received a telegram inviting her to report to Wilmington, Delaware, to be flight tested for Nancy Love's Women's Auxiliary Ferrying Squadron.[140] As it had been for Adela Scharr, Sharp's background was very different from that of most of the women reporting to the Newcastle Army Air Base. She had never experienced the way of life most of her fellow WAFS took for granted. Sharp moved into bachelor officers quarters 14 with Shanty, her Scotty dog, although animals were forbidden; but she soon realized that it was difficult to have an animal while flying around the country, so the dog did not stay. Bartels notes that Sharp's four-week training period consisted of flight training and ground school, which included standing guard, handling rifles, and dismantling and assembling a .45 caliber pistol while blindfolded.[141] One WAFS rule, a new restriction for Sharp, was the prohibition against riding in airplanes with male pilots unless ordered to do so, a rule based on the fear of public misperception. Only in officers' clubs were these "no socializing" restrictions relaxed.

"Aunt Elsie" visited her daughter while she was in the WAFS, and while the two dined in New York, she disclosed the truth about Sharp's origins.[142] Later, Sharp was stationed at Long Beach AAB in California. She transitioned to larger and faster aircraft with her fellow WAFS. Sharp discovered that she loved flying military aircraft. She spent hours in the Link trainer, learning how to fly with instruments because WAFS had not received that kind of training and she wanted to keep pace with the WASPs, who had. Sharp served as Barbara Erickson's executive officer in Long Beach.[143] When the film *Ladies Courageous* was being made, the photogenic Sharp posed for its publicity shots.[144]

While in the WASP, Sharp sent money back to Ord to help with family expenses. She was about to earn a Fifth Rating, needed for flying heavy bombers, when she checked out in a P-38. In a fighter, the first flight was of necessity a solo because there was room for only one pilot.[145] Sharp embarked on her first delivery of a P-38 from Long Beach, California, to Newark, New Jersey, in March 1944.[146] Weather conditions forced her to land in New Cumberland, Pennsylvania, rather than flying the rest of the trip in bad weather. On her takeoff from the Harrisburg airfield, her twin-engine airplane lost one engine and failed to gain altitude, crash-landing on Beacon Hill, adjacent to the airport in New Cumberland. Sharp was killed at 10:30 A.M. on April 3, 1944.[147] Nancy Batson accompanied her body back to Nebraska. A large crowd attended her funeral service in Ord. Sharp's next ferrying trip, after the one that claimed her life, would have taken her to Great Falls, Montana, where for the first time, she had planned to meet her biological father, Orla Edward Crouse.[148]

WAFS Evelyn Sharp in the cockpit. Sharp was already a well known barnstormer in Ord, Nebraska, when she joined Nancy Love's WAFS. When Sharp was killed during takeoff in a P-38, her hometown named its airport in her honor (The Woman's Collection, Texas Woman's University).

On September 12, 1948, the airfield in Ord, Nebraska, was renamed Evelyn Sharp Field, its gate graced by a war surplus P-38 propeller. Sharp's grave lies a half-mile away.[149]

Germany

Hanna Reitsch (1912–1979)

The story of Hanna Reitsch illustrates how a woman with determination and skill could work as a military pilot in wartime in a country where a woman's highest calling was thought to be raising children and keeping house. After Adolf Hitler rose to power, German attitudes toward absolutely separate sex roles only strengthened. They never relaxed, even during the war, as they did in Great Britain, the Soviet Union and the United States. Reitsch, who was born in the German town of Hirschberg in Silesia, was a phenomenon in that social context because she followed her own course, eventually becoming a test pilot for the Nazis. Reitsch was one of only six women who flew in Germany during the Second World War.[150]

Hanna Reitsch wanted to follow her father into the field of medicine, but she dreamed that her future in medicine would be as a flying doctor in Africa. Although her parents encouraged her domestic education, enrolling her in the Koloschule for that purpose, she was never inclined toward the pursuit of housekeeping, cooking, or anything similar. While a medical student, she learned to fly at the famous Grunau School for Glider Pilots, gliding being the only sanctioned form of flight allowed in Germany after the First World War and thus a highly popular sport there. During the 1930s Reitsch transitioned to powered aircraft; but gliding was, to her mind, the purest form of flight and it remained her favorite. To educate herself about airplane engines, she first learned everything she could about automobile engines. Reitsch became a stunt pilot for a movie company after she had set a gliding altitude record while buoyed by the updraft of a huge storm cloud. She was accepted into the Civil Airways Training School in Stettin, the only female student.[151] There she took part in drills along with the men, for whom the diminutive woman was an object of curiosity and interest. Her performance at the school established her credentials among recruits and instructors alike.

Usually avoiding political discussions, Reitsch noted in her autobiography that friendly relations were often spoiled when politics entered the picture. Her skill in aviation attracted attention and brought her international recognition between the wars, thanks to the German government's decision to employ her as a traveling ambassador useful for generating good will for Germany throughout the world. Reitsch's apolitical nature, in combination with her youthful candor, charmed her audiences during the 1930s. She appeared to savor her fame and she was enchanted with flight. Reitsch traveled to South America, Finland, Spain, the United States, and Africa during the interwar period. Once, after participating in an international air display in Lisbon, she wrote, "Our minds were in harmony, untouched by the tensions of the outside world, for common to us all was the one overmastering desire — to soar in the beauty of flight."[152]

Reitsch was one of several German glider pilots who visited flying clubs in England before World War II, and she participated in courses sponsored by the Anglo-German Fellowship. Ann Welch, one of the American ATA pilots, who was active in both English and German gliding clubs, met Reitsch in 1937. In her book, *Happy to Fly*, Welch commemorated the German pilot's visit to England in a cartoon titled "The Anglo-German Gliding Camp 1937."[153]

The German Luftwaffe "recruited" Reitsch in September 1937 when Ernst Udet, a former pilot in the First World War, who was then a senior Luftwaffe colonel, ordered her to report to the testing station at Rechlin. Reitsch was not warmly received by the men stationed there. No one wanted a woman on the airfield. But she characteristically persevered. Her work involved testing the dive brakes on a Stuka, a large machine for a petite pilot. Because of her service as a test pilot, Reitsch was awarded the title *Flugkapitan* (flight captain), a civilian title designed to honor the best (male) pilots.[154] Her pride in that title is obvious from the fact that she continued to use it on her stationary many years after the war ended.[155]

In 1938 Hanna Reitsch was accepted into the Luftwaffe. For several years, Germany had been violating the provisions of the Treaty of Versailles, which forbade the country from pursuing "engine-propelled" flight. During the 1938 Berlin Motor Show, Reitsch became the first woman to fly a helicopter, a new technological wonder which she demonstrated for three weeks in shows that were staged in the indoor Deutschlandhalle.[156] The program's theme was Africa, in celebration of Germany's long lost colonies. Before the Berlin display, Reitsch had flown a helicopter in Bremen for Charles Lindbergh, the visiting American aviator, who considered the machine an amazing aeronautical development. After the war, these early helicopter flights would ensure for Reitsch the honor of being named Member #1 by an international association, the Whirly Girls.[157] Reitsch's travels took her to the United States in the summer of 1938 for the International Air Races in Cleveland, Ohio. Her impression of the country was overwhelmingly positive, and she was charmed by the Americans' "unconcerned spontaneity." Reitsch would have liked to have stayed longer in the U.S., but in August she received a telegram recalling her to Germany "amid disturbing news from Czechoslovakia," she wrote later.[158]

Her autobiography covers Reitsch's activities before, during, and after the Second World War and it gives considerable insight into her thoughts and actions. She was appalled at the brutality of the Nazi regime, but at the same time, she expressed a conviction that Hitler was unaware of it. She believed that as soon as Hitler knew about the Nazis' cruelty, he would stop it. Her autobiography, *Fliegen mein Leben* (*Flying [Is] My Life*), appeared in 1951. It proves that, for Reitsch, Adolf Hitler was a man who was larger than life and a man of honor. Reitsch's preconceptions were seriously tested at her first Iron

Cross award ceremony when she discovered that her hero could be ill-mannered and uncouth, not at all the paragon she expected.[159] However, she was favorably impressed by Hitler's knowledge of aviation, which she considered "remarkable for a layman," as indicated by the "searching pointedness of his questions" on the subject.[160]

Reitsch's test-piloting assignments required her to focus almost exclusively on her day-to-day work, so her political opinions were not usually challenged in conversation with others. Claiming disinterest in politics, she considered her work as a test pilot vital to preventing the injury or death of German combat pilots. Her test-piloting work sought out whatever flaws existed in German aircraft and it pushed their technological limits. Reitsch conducted trials of experimental and unusual craft, such as the Gigant, an enormous glider that was nearly impossible to haul off the ground. She gauged the stability of a gasoline tanker designed to be towed behind a pilotless glider. She also tested a cutting device for barrage balloons, which were used widely in England and Russia to protect cities, factories, and airbases, and which constituted a serious hazard to German pilots.

During a 1942 test of the Messerschmitt Me 163 rocket plane, which reached speeds of 500 miles per hour, the jet stalled, and Reitsch, who opted not to bail out but to land the airplane, was seriously injured in a crash landing. Her recovery from that accident took many months. The German government rewarded her valor by presenting her an Iron Cross, First Class, and a military gold medal with diamonds. Reitsch was not allowed to return to the Me 163 after she'd recovered because she'd lost so much flight time while recuperating, so she was sent to the Russian front in 1943, an attempt to boost the flagging morale of German soldiers. However, upon seeing their poor condition, Reitsch herself became demoralized.[161]

Toward the end of the war, Hanna Reitsch was convinced that a piloted flying bomb was the only remedy to Germany's eroding chance of victory, and she volunteered for one of these suicide missions. At the presentation of her second Iron Cross (Iron Cross, First Class), and by then certain that Hitler was "living in some remote and nebulous world of his own," she tried to convince the fuehrer that suicide missions were imperative, but the plans never materialized.[162] Her biographers think it is most likely that Reitsch was unaware of the Nazi atrocities. When she was shown photographs from the death camps at war's end, she appeared unable to believe that they were real.

In April 1945, Reitsch flew to Hitler's bunker in Soviet-occupied Berlin with her friend, General Ritter von Greim, who had been ordered to report to the chancellery. Von Greim flew the first leg of their extremely dangerous journey in a Focke-Wulf 190 and landed the airplane in Gatow, the only Berlin airfield still under German control. There he and Reitsch exchanged their first airplane for a Fieseler Storch. Along their route, von Griem's foot was hit by Russian fire from below, so Reitsch continued as pilot, controlling the airplane while leaning over her injured friend's shoulder. After the couple arrived in the bunker, Reitsch was allowed to telephone her family in Salzburg. She described her flight to Berlin in such detail that her narrative ceased only after an impatient general cut the connection to keep the line open for more pressing matters. After a few days von Greim and Reitsch reluctantly departed, carrying letters from the bunker's inhabitants, although she admitted later that she tore up Eva Braun's letter to her sister, believing it vulgar and theatrical. Reitsch also held poison vials for her and von Greim, parting gifts from Hitler.[163]

Hanna Reitsch in her uniform and medals in 1940. Reitsch was proud of her test-piloting work during the war because it made flying safer for German pilots (The Woman's Collection, Texas Woman's University).

Knowing that the war was a lost cause for Germany, deeply afraid of the Soviets, and unaware of his daughter's fate, Reitsch's father shot the five members of his household and then shot himself.[164] Only Hanna and her brother, Kurt, survived the war and eventually reunited. Reitsch was interviewed by Robert Work, a British journalist, in October 1945. She was one of seven witnesses questioned about Hitler's last days because of the time she had spent in his bunker from April 26 to April 29. Some of her testimony could not be corroborated by other witnesses, so it was considered unreliable. Some of the statements she made at that time were later proved false by other witnesses. Suspected of spiriting Hitler away from Berlin on her return flight, Reitsch was interrogated by the American Counter Intelligence Corps (CIC). She was detained for fifteen months until August 1946, and for a while she was under surveillance by the CIC, which erroneously assumed she was a Nazi.[165]

For the rest of her life, Hanna Reitsch claimed to have felt unfairly vilified as a German and she rued the rumors associating her with Hitler. Ineligible for a military pension, she lived meagerly in Germany during the postwar years. Eventually she became active in reviving the gliding movement in her country, but she sparred with the German Aero Club over her controversial past. She travelled to India during the 1950s and returned to Finland and the United States in 1960. Kwam Nkrumah invited her to set up a gliding school in Ghana, and she worked in Africa until Nkrumah's overthrow in 1966. She was (along with America's Jacqueline Cochran and France's Jacqueline Auriol) an honorary member of the Society of Experimental Test Pilots. Reitsch was honored as "Pilot of the Year" in Arizona, and she received the International Award of Characters under the name "Supersonic Sue."[166] Reitsch was sure that aviation, specifically gliding, could promote peace and understanding among nations. She ended her autobiography with the following words: "I long to fly again! I long to be soaring up into those timeless and hallowed spaces, whose beauty compels all true airmen to a belief in God. May all we who live to fly show forth that belief to our fellow men! And may gliding in future be allowed only to bless mankind, for what better instrument could there be than that lovely experience to bring peace and reconciliation to the peoples of the world?"[167]

Hanna Reitsch died in Bonn, Germany on August 26, 1979, at the age of sixty-seven. Her death was announced to the public only after her private burial was over.

Melitta Schiller, Countess von Stauffenberg (1903–1945)

Melitta Schiller was born in Prussia, and although her father had converted to Protestantism, her paternal family's heritage was Jewish. In 1937, Schiller married Alexander Schenk Graf von Stauffenberg, the brother of Claus, who would launch an unsuccessful assassination attempt against Adolf Hitler. Before the war, Schiller became the second woman to be awarded the title of *Flugkapitan*. She and the other female flight captain, Hanna Reitsch, were "bitter rivals" according to Lomax.[168] When the war began, Schiller became a test pilot, flying "well over two thousand diving missions in 'Ju[nkers] 87' and 'Ju 88' dive-bombers, an accomplishment that was surpassed by only one other German pilot, a man." Schiller's Jewish heritage would have been more dangerous to her in wartime Germany had it not been for her high value to the Nazis as a pilot.[169] As Countess von Stauffenberg, she was placed in charge of the test station for special flying equipment at the Air War Academy at Berlin-Gatow.[170] She performed night flights with the Arado 96, the Focke-Wulf 190, and the turbo-jet fighter Messerschmitt 262, and she worked on night landing instruments. After her husband and other members of the von Stauffenberg family were imprisoned for their participation in the failed assassination attempt, Schiller took advantage of her "war-essential" status to deliver food to him and other family members. She obtained permission to make flights at her discretion, flying a slow Fieseler Storch on her delivery missions. She visited the camp at Buchenwald at least twice. On April 8, 1945, Schiller's plane, a slow, unarmed Bucker 181 trainer, was shot down by an American while she flew near Strasskirchen. Schiller landed the airplane successfully, but afterwards she died of her wounds.[171]

Chapter Notes

Preface

1. K. Jean Cottam, e-mail message to author, February 27, 2008.
2. Peter Hoffmann, *Stauffenberg: A Family History, 1905–1944* (Montreal & Kingston: McGill-Queen's University Press, 1995), 280.

Introduction

1. Nancy Bird, *My God! It's a Woman* (Sydney, Australia: Angus & Robertson, 1990), 153.

Chapter 1

1. "Stinson, Katherine," *Handbook of Texas Online*, http://www.tshaonline.org/handobok/online/articles/SS/fst97 (accessed March 11, 2008).
2. "Aviatrices — Ruth Law," *Hargrave the Pioneers*, http://www.ctie.monash.edu.au/hargrave/pioneers (accessed January 21, 2010).
3. Michael Fahie, *Harvest of Memories: The Life of Pauline Gower, M.B.E.* (Peterborough, UK: GMS, 1995), 87.
4. Pauline Gower, *Women with Wings* (London: John Long, 1938), 97.
5. Ibid., 77.
6. Ibid., 74.
7. Kazimiera Janina Cottam, *Women in War and Resistance: Selected Biographies of Soviet Women Soldiers* (Nepean, Canada: New Military, 1998), 23.
8. Ibid., 4.
9. Doris Rich, *Jackie Cochran: Pilot in the Fastest Lane* (Gainesville: University Press of Florida Press, 2007), 29.
10. Leslie Haynsworth and David Toomey, *Amelia Earhart's Daughters: The Wild and Glorious Story of American Women Aviators from World War II to the Dawn of the Space Age* (New York: William Morrow, 1998), 19.
11. Sally Knapp, *New Wings for Women* (New York: Thomas Y. Crowell, 1946), 52.
12. "Nancy Harkness Love," National Aviation Hall of Fame, http://nationalaviation.blade6.conet.com/components/content (accessed March 15, 2008).
13. Giles Whittell, *Spitfire Women of World War II* (London: HarperPress, 2007), 67.
14. Joseph Stalin, *Great Patriotic War of the Soviet Union* (New York: International, 1945), 11.
15. Cottam, *Women in War*, 41.
16. Dominick A. Pisano, *To Fill the Skies with Pilots: The Civilian Pilot Training Program, 1939–46* (Urbana and Chicago: University of Chicago Press, 1993), 61.
17. Ibid., 76.
18. Ibid., 77.

Chapter 2

1. Fahie, *Harvest of Memories*, 135.
2. Ibid., 138.
3. E.C. Cheesman, *Brief Glory: The Story of A.T.A.* (Leicester, UK: Harborough, 1946), 12.
4. Shirley Render, *No Place for a Lady: The Story of Canadian Women Pilots, 1928–1992* (Winnipeg, Manitoba, Canada: Portage & Main, 1992), 81.
5. Fahie, *Harvest of Memories*, 141.
6. Ibid., 141.
7. Cheesman, *Brief Glory*, 73.
8. Kazimiera Janina Cottam, *Women in Air War: The Eastern Front of World War II* (Nepean, Canada: New Military, 1997), 6.
9. Reina Pennington, *Wings, Women and War: Soviet Airwomen in World War II Combat* (Lawrence: University Press of Kansas, 2001), 57.
10. Ibid., 47.
11. Rich, *Jackie Cochran*, 99.
12. Ibid., 106.
13. H.H. Arnold, *Global Mission* (New York: Arno, 1949), 311.
14. Rich, *Jackie Cochran*, 110.
15. Whittell, *Spitfire Women*, 138.
16. Fahie, *Harvest of Memories*, 174.

17. Whittell, *Spitfire Women*, 207.
18. Sarah Byrn Rickman, *Nancy Love and the WASP Ferry Pilots of World War II* (Denton: University of North Texas, 2008), 62.
19. Ibid., 65.
20. Ibid., 76.
21. Ibid., 80.
22. Betty Stagg Turner, *Out of the Blue and into History* (Arlington Heights, IL: Aviatrix, 2001), 12.
23. Rich, *Jackie Cochran*, 129.
24. Ibid., 128.

Chapter 3

1. Jennet Conant, *The Irregulars: Roald Dahl and the British Spy Ring in Wartime Washington* (New York: Simon & Schuster, 2008), 42–43.
2. Rosemary du Cros, *ATA Girl: Memoirs of a Wartime Ferry Pilot* (London: Frederick Muller, 1983), 61.
3. Veronica Volkersz, *The Sky and I* (London: W.H. Allen, 1956), 35.
4. Beryl E. Escott, *The WAAF: A History of the Women's Auxiliary Air Force in the Second World War* (Buckinghamshire, UK: Shire, 2003), 18.
5. Beryl E. Escott, *Women in Air Force Blue: The Story of Women in the Royal Air Force from 1918 to the Present Day* (Northamptonshire, UK: Patrick Stephens, 1989), 178.
6. Volkersz, *The Sky and I*, 17.
7. Elizabeth Strohfus and Cheryl J. Young, *Love at First Flight: One Woman's Experience as a WASP in World War II ... and Fifty Years Later, She's Still Flying* (St. Cloud, MN: North Star Press, 1994), 14.
8. Anna Timofeyeva-Yegorova, *Red Sky, Black Death: A Soviet Woman Pilot's Memoir of the Eastern Front* (Bloomington, IN: Slavica, 2009), 35.
9. Pennington, *Wings, Women and War*, 25.
10. Kazimiera Janina Cottam, *In the Sky above the Front: A Collection of Memoirs of Soviet Airwomen, Participants in the Great Patriotic War* (Manhattan, KS: MA/AH, 1984), 37.
11. Knapp, *New Wings*, 139.
12. O. Vivian Fagan, *Zoot Suits and Parachutes: WASPs WWII* (Hawaii[?]: O.V. Fagan, 1999), 14.
13. Rob Simbeck, *Daughter of the Air: The Brief Soaring Life of Cornelia Fort* (New York: Atlantic Monthly Press, 1999), 126.
14. Diane Ruth Armour Bartels, *Sharpie: The Life Story of Evelyn Sharp, Nebraska's Aviatrix* (Lincoln, NE: Dageforde, 1996), 171.
15. Ibid., 182.
16. Pennington, *Wings, Women and War*, 32.
17. Whittell, *Spitfire Women*, 65.
18. Fahie, *Harvest of Memories*, 141.
19. Ibid., 171.
20. Diane Barnato Walker, *Spreading My Wings: One of Britain's Top Woman Pilots Tells Her Remarkable Story* (Somerset, UK: Patrick Stephens, 1994), 45.
21. Fahie, *Harvest of Memories*, 175.
22. Lettice Curtis, *The Forgotten Pilots: A Story of the Air Transport Auxiliary 1939–45* (Henley-on-Thames, Oxfordshire, UK: G.T. Foulis, 1971), 80.
23. Yegorova, *Red Sky*, 2.
24. Pennington, *Wings, Women and War*, 31–34.
25. Ibid., 32.
26. "Requirements," WASP on the Web, http://sss.wingsacrossamerica.us (accessed October 24, 2008).
27. Amy Goodpaster Strebe, *Flying for Her Country: The American and Soviet Women Military Pilots of World War II* (Westport, CT: Praeger Security International, 2007), 8.
28. "Women's Airforce flight training," WASP on the Web, http://www.wingsacrossamerica.us (accessed October 24, 2008).
29. Olga Gruhzit-Hoyt, *They Also Served: American Women in World War II* (New York: Birch Lane, 1995), 13.
30. Walker, *Spreading My Wings*, 50.
31. Render, *No Place for a Lady*, 79.
32. Ibid., 87.
33. Ibid., 121.
34. Joyce A. Thomson, *The WAAAF in Wartime Australia* (Carlton, Victoria, Australia: Melbourne University Press, 1991), 41.
35. Ibid., 347.
36. Doris Ferry, *Western Memories: History and Recollections of the Western Australian WAAAF 1941–1946* (Woodvale, Western Australia: Doris Ferry, 1996), 19.
37. Clare Stevenson and Honor Darling, *The WAAAF Book* (Sidney, NSW, Australia: Hale & Iremonger, 1984), 133.
38. Ibid., 164.
39. Bird, *My God! It's a Woman*, 160.
40. Constance Babington Smith, *Amy Johnson* (London: Collins, 1967), 139.
41. Delores Theresa Moggridge, *Woman Pilot* (London: Michael Joseph), 1957, 64.
42. Cheesman, *Brief Glory*, 78.
43. Knapp, *New Wings*, 90.
44. Curtis, *Forgotten Pilots*, 312.
45. Whittell, *Spitfire Women*, 118.
46. Ibid., 75.
47. Ibid., 118–124.
48. Regina Trice Hawkins, *Hazel Jane Raines: Pioneer Lady of Flight* (Macon, GA: Mercer University Press, 1996), 92.
49. Ibid., 6.
50. "Americans in the British ATA," MSS 392c, Woman's Collection, Blagg-Huey Library, Texas Woman's University, Denton.

51. Whittell, *Spitfire Women*, 138.
52. "Americans in the British ATA," MSS 392c, TWU.
53. Whittell, *Spitfire Women*, 155.
54. Gower, *Women with Wings*, 72.
55. Ibid., 116.
56. Lazar Brontman and L. Khvat, *The Heroic Flight of the "Rodina"* (Moscow: Foreign Languages, 1938), 94.
57. Pennington, *Wings, Women and War*, 31.
58. Von Hardesty and Dominick Pisano, *Black Wings: The American Black in Aviation* (Washington, D.C.: National Air and Space Museum, 1983), 17.
59. Ibid., 21.
60. Ibid., 57.
61. Jacqueline Cochran, *The Stars at Noon* (Boston: Little, Brown, 1954), 128.
62. "African American WASP Applicants," MSS 250, Blagg-Huey Library, Texas Woman's University, Denton.
63. Janet Harmon Bragg, *Soaring Above Setbacks: The Autobiography of Janet Harmon Bragg, African American Aviator* (Washington, D.C.: Smithsonian Institution Press, 1996), 40.
64. "African American WASP applicants," MSS 250, TWU.
65. Adela Riek Scharr, *Sisters in the Sky*, vol. 2, *The WASP* (St. Louis, MO: Patrice, 1986), 636.
66. Ibid., 672.
67. Haynsworth and Toomey, *Amelia Earhart's Daughters*, 110.
68. Turner, *Out of the Blue*, 83.
69. Wayne Reese, Alan H. Rosenburg, and Ming-Na, *A Brief Flight: Hazel Ying Lee and the Women Who Flew Pursuit*, VHS (Pacific Palisades, CA: LAWAS Productions, 2002).
70. "WASP biofiles — 44-W-9 — Gee, Margaret (Maggie)," MSS 250, TWU.
71. Pennington, *Wings, Women and War*, 31.
72. Sally Van Wagenen Keil, *Those Wonderful Women in Their Flying Machines: The Unknown Heroines of World War II* (New York: Rawson, Wade), 114.
73. Rich, *Jackie Cochran*, 119.
74. Bartels, *Sharpie*, 219.
75. Strohfus and Young, *Love at First Flight*, 50.
76. Marie Mountain Clark, *Dear Mother and Daddy: World War II Letters Home from a WASP, an Autobiography* (Livonia, MI: First Page, 2005), 97.
77. Bartels, *Sharpie*, 220.
78. Byrd Howell Granger, *On Final Approach: The Women Airforce Service Pilots of W.W.II* (Scottsdale, AZ: Falconer, 1991), 108.
79. *WASPs and Witches: The History of Women Pilots Fighting for the Right to Fight*, DVD, produced and directed by Jamie Doran (Atlantic Celtic Films, 2000).
80. Gruhzit-Hoyt, *They Also Served*, 168.
81. Fagan, *Zoot Suits and Parachutes*, 78.
82. Bartels, *Sharpie*, 250.
83. Turner, *Out of the Blue*, 275.
84. Strohfus and Young, *Love at First Flight*, 32.
85. Clark, *Dear Mother and Daddy*, 72.
86. Ken Magid, *Women of Courage: The Story of the Women Pilots of World War II*, VHS (Lakewood, CO: K.M. Productions, 1993).
87. Fagan, *Zoot Suits and Parachutes*, 123.
88. Adela Riek Scharr. *Sisters in the Sky*, vol. 1, *The WAFS* (St. Louis, MO: Patrice, 1986), 290.
89. Ibid., 519.
90. Fahie, *Harvest of Memories*, 141.
91. Ibid., 178.
92. Volkersz, *The Sky and I*, 68.
93. Ibid., 99–100.
94. Glenn Kerfoot, *Propeller Annie: The Story of Helen Richey, the Real First Lady of the Airlines* (Lexington: Kentucky Aviation History Roundtable, 1988), 85.
95. *Daily Life Online*, "Daily Life through History and World Cultures Today" (Westport, CT: Greenwood, c2004), http://0-dailylife.greenwood.com "daily life in the Soviet Union" — "Soviet constitutions" (accessed December 10, 2008).
96. Pennington, *Wings, Women and War*, 66.
97. *Daily Life Online*, "Daily Life in the Soviet Union," "The Military," "Deferments," "Officers," "Soldier's Daily Life, 1922–1939," http://0-dailylife.greenwood.com (accessed December 10, 2008).
98. Jan Churchill, *On Wings to War: Teresa James, Aviator* (Manhattan, KS: Sunflower University Press, 1992), 59.
99. Doris Brinker Tanner, *Zoot-Suits and Parachutes and Wings of Silver Too!: The World War II Air Force Training of Women Pilots 1942–1944* (Paducah, KY: Turner, 1996), 18.
100. WASP on the Web, "Officer's pay comparison," http://www.wingsacrossamerica.us (accessed October 24, 2008).
101. Curtis, *Forgotten Pilots*, 308.
102. Pennington, *Wings, Women and War*, 234.
103. Turner, *Out of the Blue*, 12.
104. Kay Gott, *Women in Pursuit: Flying Fighters for the Air Transport Command Ferrying Division during World War II, a Collection & Recollection* (McKinleyville, CA: Kay Gott, 1993), 11.
105. WASP on the Web, "WASP facts," http://www.wingsacrossamerica.us (accessed January 19, 2010).
106. Fahie, *Harvest of Memories*, 145.
107. du Cros, *ATA Girl*, 36, 40.
108. Ibid., 37–38.

109. Strebe, *Flying for Her Country*, 43.
110. Pennington, *Wings, Women and War*, 63.
111. Stalin, *Great Patriotic War*, 34.
112. Haynsworth and Toomey, *Amelia Earhart's Daughters*, 60.
113. Tanner, *Zoot-Suits*, 16.
114. Render, *No Place for a Lady*, 79.
115. Marianne Verges, *On Silver Wings: The Women Airforce Service Pilots of World War II 1942–1944* (New York: Ballantine, 1991), 87.
116. Yvonne Pateman and Gene Wyall, *We Were WASP: Women Airforce Service Pilots, WWII*, VHS (Mission Viejo, CA: Video Movie Magic, 1990).

Chapter 4

1. Curtis, *Forgotten Pilots*, 41.
2. Volkersz, *The Sky and I*, 29.
3. Ibid., 37.
4. Walker, *Spreading My Wings*, 50.
5. Ibid.
6. "Americans in the British ATA," MSS 392c, TWU.
7. Vi Warren Milstead, "Flying with the ATA,"*CAHS Journal* (Fall 1999): 111.
8. Pennington, *Wings, Women and War*, 31.
9. Ibid., 40.
10. Ibid., 42.
11. Ibid., 47.
12. Cottam, *In the Sky*, 48.
13. Pennington, *Wings, Women and War*, 44.
14. Ibid., 30.
15. Ibid., 42.
16. Helene Keyssar and Vladimir Pozner, *Remembering War: A U.S.–Soviet Dialogue* (New York: Oxford University Press, 1990), 43.
17. Cottam, *In the Sky*, 180.
18. Pennington, *Wings, Women and War*, 42.
19. Ibid., 48.
20. Tanner, *Zoot-Suits*, 14.
21. Simbeck, *Daughter of the Air*, 134.
22. Churchill, *On Wings to War*, 76.
23. Tanner, *Zoot-Suits*, 33.
24. Clark, *Dear Mother and Daddy*, 79.
25. Strohfus and Young, *Love at First Flight*, 33.
26. Tanner, *Zoot-Suits*, 29–30.
27. Ann L. Cooper, *How High She Flies: Dorothy Swain Lewis, WASP of WWII, Horsewoman, Artist, Teacher* (Arlington Heights, IL: Aviatrix, 1999), 68.
28. "Aviatrices—Phoebe Omlie," *Hargrave the Pioneers*, http://www.ctie.monash.edu.au/hargrave/pioneers (accessed January 21, 2010).
29. Cooper, *How High She Flies*, 70.
30. Ibid., 75.
31. Ibid., 81.
32. Ibid., 84.
33. Fagan, *Zoot Suits and Parachutes*, 93.
34. Keil, *Those Wonderful Women*, 351–352.
35. Clark, *Dear Mother and Daddy*, 146.
36. Haynsworth and Toomey, *Amelia Earhart's Daughters*, 63.
37. Whittell, *Spitfire Women*, 17–18.
38. Strohfus and Young, *Love at First Flight*, 49.
39. Tanner, *Zoot-Suits*, 97–98.
40. Ibid., 41.
41. Clark, *Dear Mother and Daddy*, 58.
42. Ibid., 62.
43. Tanner, *Zoot-Suits*, 47–48.
44. WASP on the Web, "WASP resources, On the record 1940s, 1944 WASP newsletters," http://www.wingsacrossamerica.us (accessed January 23, 2010).
45. Fagan, *Zoot Suits and Parachutes*, 58.
46. Ibid., 103.
47. Milstead, "Flying with the ATA,"*CAHS Journal* (Fall 1999): 88–89.
48. Curtis, *Forgotten Pilots*, 43–44.
49. Knapp, *New Wings for Women*, 97.
50. Pennington, *Wings, Women and War*, 39.
51. Ibid., 2.
52. Ibid., 46.
53. Ibid., 122.
54. Churchill, *On Wings to War*, 53–55.
55. Ibid., 53.
56. Tanner, *Zoot-Suits*, 27.
57. Clark, *Dear Mother and Daddy*, 29.
58. Tanner, *Zoot-Suits*, 33.
59. Verges, *On Silver Wings*, 79.
60. WASP on the Web, "Uniforms," http://www.wingsacrossamerica.us (accessed July 27, 2009).
61. Granger, *On Final Approach*, 125.
62. Betty Greene and Dietrich Buss, *Flying High: The Amazing Story of Betty Greene and the Early Years of Mission Aviation Fellowship* (Camp Hill, PA: Christian, 2002), 36.
63. WASP on the Web, "Timeline of WASP history," http://www.wingsacrossamerica.us (accessed January 28, 2008).
64. Conant, *Irregulars*, 42.
65. Gower, *Women with Wings*, 200.
66. Conant, *Irregulars*, 45.
67. Fahie, *Harvest of Memories*, 150.
68. Curtis, *Forgotten Pilots*, 198.
69. du Cros, *ATA Girl*, 54–55.
70. Ibid., 83.
71. Ibid., 81.
72. *WASPs and Witches*, DVD.
73. Cheesman, *Brief Glory*, 56.
74. Volkersz, *The Sky and I*, 57.
75. Ibid., 93–94.
76. Ibid., 94.
77. Ibid., 87–88.
78. Cottam, *In the Sky*, 87.
79. Ibid., 236.
80. Ibid., 238.

81. Ibid., 94–95.
82. Ibid., 45.
83. Ibid., 26.
84. Keyssar and Pozner, *Remembering War*, 40.
85. Cottam, *In the Sky*, 48.
86. Ibid., 226.
87. Knapp, *New Wings for Women*, 138.
88. Cottam, *In the Sky*, 226.
89. *WASPs and Witches*, DVD.
90. Strebe, *Flying for Her Country*, 18.
91. Granger, *On Final Approach*, 278.
92. Adams, Nelson, et al., *Silver wings and Santiago blue*, VHS.
93. Clark, *Dear Mother and Daddy*, 112.
94. Simbeck, *Daughter of the Air*, 192.
95. Pennington, *Wings, Women and War*, 67.
96. Verges, *On Silver Wings*, 169.
97. Margaret J. Ringenberg and Jane L. Roth, *Girls Can't Be Pilots: An Aerobiography* (Fort Wayne, IN: Daedalus, 1998), 75.
98. Verges, *On Silver Wings*, 169.
99. Gruhzit-Hoyt, *They Also Served*, 161.
100. Tanner, *Zoot-Suits*, 42.
101. Clark, *Dear Mother and Daddy*, 21–22.
102. Scharr, *Sisters in the Sky*, vol. 2, p. 69.
103. Fagan, *Zoot Suits and Parachutes*, 137.
104. Scharr, *Sisters in the Sky*, vol. 2, p. 305.
105. Strohfus and Young, *Love at First Flight*, 41–45.
106. Whittell, *Spitfire Women*, 44.
107. Pennington, *Wings, Women and War*, 106.
108. Ibid., 111.
109. Ibid., 109.
110. Ibid.
111. Ibid.
112. Pennington, *Wings, Women and War*, 113.
113. Granger, *On Final Approach*, 33.
114. Ibid., 104.
115. Rich, *Jackie Cochran*, 128.
116. Gott, *Women in Pursuit*, 34.
117. Clark, *Dear Mother and Daddy*, 85.
118. Rich, *Jackie Cochran*, 106.
119. Scharr, *Sisters in the Sky*, vol. 2, p. 66.
120. Bird, *My God! It's a Woman*, 145.
121. Jack Fisher and Robert O'Gorman, *Women Combat Pilots: The Right Stuff*, DVD (New York: A & E Television Networks, 2003).
122. Bartels, *Sharpie*, 220.
123. Rich, *Jackie Cochran*, 115.

Chapter 5

1. Cheesman, *Brief Glory*, 152.
2. Curtis, *Lettice Curtis*, 71.
3. Walker, *Spreading My Wings*, 51–52.
4. Hawkins, *Hazel Jane Raines*, 93.
5. WASP on the Web, "WASP timeline," http://www.wingsacrossamerica.us (accessed January 28, 2008).
6. Rickman, *Nancy Love*, 106–107.
7. Granger, *On Final Approach*, 124.
8. Ringenberg and Roth, *Girls Can't Be Pilots*, 8.
9. Clark, *Dear Mother and Daddy*, 111.
10. Turner, *Out of the Blue*, 455.
11. WASP on the Web, "Bases," http://www.wingsacrossamerica.us (accessed January 28, 2008).
12. Ann Baumgartner Carl, *A WASP Among Eagles: A Woman Military Test Pilot in World War II* (Washington, D.C.: Smithsonian Institution Press, 1999), 111.
13. Gruhzit-Hoyt, *They Also Served*, 157.
14. Fagan, *Zoot Suits and Parachutes*, 114.
15. Fahie, *Harvest of Memories*, 142.
16. Ibid., 146.
17. Ibid., 153.
18. Ibid., 145–146.
19. Ibid., 167.
20. Ibid., 174.
21. Ibid., 180.
22. Ibid., 173.
23. Rickman, *Nancy Love*, 99.
24. Ibid., 108.
25. Ibid., 109.
26. Ibid., 118–119.
27. "Americans in the British ATA," MS 392c, TWU.
28. Cooper, *How High She Flies*, 71.
29. Keil, *Those Wonderful Women*, 351.
30. Scharr, *Sisters in the Sky*, vol. 2, p. 123.
31. du Cros, *ATA Girl*, 58.
32. Fahie, *Harvest of Memories*, 3.
33. Hawkins, *Hazel Jane Raines*, 23.
34. Render, *No Place for a Lady*, 89.
35. Walker, *Spreading My Wings*, 42.
36. Fahie, *Harvest of Memories*, 149.
37. Milstead, "Flying with the ATA," *CAHS Journal* (Fall 1999): 87.
38. "Americans in the British ATA," MS 392c, TWU.
39. du Cros, *ATA Girl*, 84.
40. Fahie, *Harvest of Memories*, 172.
41. du Cros, *ATA Girl*, 77–78.
42. "Americans in the British ATA," MS 392c, TWU.
43. du Cros, *ATA Girl*, 48–49.
44. Knapp, *New Wings for Women*, 97.
45. du Cros, *ATA Girl*, 50.
46. Milstead, "Flying with the ATA," 92.
47. "Americans in the British ATA," MS 392c, TWU.
48. Whittell, *Spitfire Women*, 238.
49. Hawkins, *Hazel Jane Raines*, 93.
50. Milstead, "Flying with the ATA," *CAHS Journal* (Fall 1999): 84.
51. Walker, *Spreading My Wings*, 53.

52. "Americans in the British ATA," MS 392c, TWU.
53. Volkersz, *The Sky and I*, 41.
54. "Americans in the British ATA," MS 392c, TWU.
55. Curtis, *Forgotten Pilots*, 73.
56. Milstead, "Flying with the ATA," *CAHS Journal* (Fall 1999): 91–92.
57. Whittell, *Spitfire Women*, 256.
58. Ibid., 259.
59. du Cros, *ATA Girl*, 71–72.
60. Keil, *Those Wonderful Women*, 379.
61. Gott, *Hazel Ah Ying Lee*, 90–92.
62. Ibid., 58.
63. Ibid., 59.
64. Strohfus and Young, *Love at First Flight*, 41.
65. Rickman, *Nancy Love*, 106–107.
66. Churchill, *On Wings to War*, 125.
67. Kerfoot, *Propeller Annie*, 92.
68. Whittell, *Spitfire Women*, 207.
69. Ibid., 271.
70. Greene and Buss, *Flying High*, 14.
71. Yvonne C. Pateman, *Women Who Dared: American Female Test Pilots, Flight-test Engineers, and Astronauts, 1912–1996* (Norstahr, 1997), 31.
72. Carl, *WASP Among Eagles*, 52–53.
73. Greene and Buss, *Flying High*, 15–18.
74. Ibid., 18, 24.
75. Ibid., 25.
76. Carl, *WASP Among Eagles*, 97.
77. Scharr, *Sisters in the Sky*, vol. 2, p. 480.
78. Cooper, *How High She Flies*, 86.
79. Ibid., 86–87.
80. Clark, *Dear Mother and Daddy*, 153.
81. Ibid., 111.
82. Granger, *On Final Approach*, 262; Turner, *Out of the Blue*, 241.
83. Strohfus and Young, *Love at First Flight*, 55.
84. Arnold, *Global Mission*, 522.
85. Churchill, *On Wings to War*, 125.
86. Render, *No Place for a Lady*, 89.

Chapter 6

1. Granger, *On Final Approach*, 438.
2. Magid, *Women of Courage*, VHS.
3. Haynsworth and Toomey, *Amelia Earhart's Daughters*, 99.
4. Laurel Ladevich, Mary McDonnell and David G. McCullough, *American Experience: Fly Girls*, DVD (Boston: WGBH, 2006), DVD.
5. Rich, *Jackie Cochran*, 132.
6. Scharr, *Sisters in the Sky*, vol. 1, p. 460.
7. Ibid., 482.
8. Ibid., 483.
9. Render, *No Place for a Lady*, 107.
10. Churchill, *On Wings to War*, 95.
11. Gruhzit-Hoyt, *They Also Served*, 173–174.
12. Volkersz, *The Sky and I*, 39.
13. Whittell, *Spitfire Women*, 210–213.
14. Walker, *Spreading My Wings*, 74.
15. Ibid., 77.
16. Ibid.
17. Walker, *Spreading My Wings*, 74.
18. Keil, *Those Wonderful Women*, 383.
19. Kerfoot, *Propeller Annie*, 86.
20. Ann R. Johnson, "The WASP of World War II," *Aerospace Historian* 17, no. 2–3 (1970): 76–82.
21. Volkhersz, *The Sky and I*, 39.
22. Walker, *Spreading My Wings*, 63.
23. Curtis, *Lettice Curtis*, 56.
24. Render, *No Place for a Lady*, 81.
25. Clark, *Dear Mother and Daddy*, 43–44.
26. Johnson Space Center Oral History Project, "Florene Miller Watson," http://www.jsc.nasa.gov/history/oralhistories/aviatrix.htm (accessed November 15, 2009).
27. Magid, *Women of Courage*, VHS.
28. Fagan, *Zoot Suits and Parachutes*, 78.
29. Tanner, *Zoot-Suits*, 118–120.
30. Jean Hascall Cole, *Women Pilots of World War II* (Salt Lake City: University of Utah Press, 1992), 116.
31. Fagan, *Zoot Suits and Parachutes*, 153–154.
32. Ibid., 152–153.
33. Carl, *WASP Among Eagles*, 51.
34. Rich, *Jackie Cochran*, 91.
35. Carl, *WASP Among Eagles*, 51
36. Greene and Buss, *Flying High*, 14.
37. Carl, *WASP Among Eagles*, 54.
38. Ibid., 61.
39. Granger, *On Final Approach*, 266.
40. Ibid., 267.
41. Pennington, *Wings, Women and War*, 74, 92, 106.
42. Cottam, *In the Sky*, 69.
43. Pennington, *Wings, Women and War*, 116.
44. Ibid., 10.
45. Ibid., 33.
46. Ibid., 104, 125.
47. Ibid., 122.
48. Ibid., 189.
49. Ibid., 125.
50. Strebe, *Flying for Her Country*, 45.
51. Cottam, *Women in Air War*, 20.
52. Pennington, *Wings, Women and War*, 90.
53. Ibid., 72.
54. Cottam, *Women in Air War*, 20.
55. Keyssar and Pozner, *Remembering War*, 38.
56. Ibid., 42.
57. Cottam, *In the Sky*, 125.
58. Cottam, *Women in Air War*, 172.
59. Cottam, *Women in War*, 43.
60. Cottam, *In the Sky*, 112.

61. Magid, *Women of Courage*, VHS.
62. Pennington, *Wings, Women and War*, 51.
63. *WASPs & Witches*, DVD.
64. *Amazons to Fighter Pilots: A Biographical Dictionary of Military Women* (Westport, CT: Greenwood Press, 2003), s.v. "Raskova" (by Reina Pennington), 353.
65. Pennington, *Wings, Women and War*, 90.
66. Cottam, *In the Sky*, 240.
67. Strebe, *Flying for Her Country*, 46–47.
68. Yegorova, *Red Sky*, 115.
69. Ibid., 183, 209.
70. Ibid., 185.
71. Ibid., 186.
72. Ibid., 108, 112.
73. Ibid., 206.
74. Ibid., 210.
75. Cottam, *Women in War*, 33.
76. Pennington, *Wings, Women and War*, 129.
77. Curtis, *Forgotten Pilots*, 309.
78. du Cros, *ATA Girl*, 65.
79. Fahie, *Harvest of Memories*, 153.
80. Whittell, *Spitfire Women*, 214.
81. Ibid., 216.
82. Yegorova, *Red Sky*, xx.
83. Ibid.
84. Keyssar and Pozner, *Remembering War*, 41.
85. Pennington, *Wings, Women and War*, 177–192.
86. Ringenberg, *Girls Can't Be Pilots*, 65.
87. Pateman, *Women Who Dared*, 45, 130.
88. Langley, *Flying Higher*, 103.
89. Verges, *On Silver Wings*, 229.

Chapter 7

1. Knapp, *New Wings for Women*, 59.
2. Ibid., 60.
3. Cochran, *Stars at Noon*, 130.
4. Cottam, *Women in War*, 25.
5. Knapp, *New Wings for Women*, 32.
6. Ibid., 33.
7. Granger, *On Final Approach*, 339.
8. Verges, *On Silver Wings*, 202.
9. Ibid., 192–193.
10. Rich, *Jackie Cochran*, 135.
11. Verges, *On Silver Wings*, 208–209.
12. Fisher and O'Gorman, *Women Combat Pilots*, DVD.
13. Verges, *On Silver Wings*, 107.
14. Rich, *Jackie Cochran*, 120.
15. Maryann Bucknam Brinley, *Jackie Cochran: An Autobiography* (New York: Bantam, 1987), 211.
16. Rich, *Jackie Cochran*, 139.
17. Granger, *On Final Approach*, 345.
18. Clark, *Dear Mother and Daddy*, 160–161.
19. Strohfus and Young, *Love at First Flight*, 38.
20. Scharr, *Sisters in the Sky*, vol. 2, p. 536.
21. Clark, *Dear Mother and Daddy*, 149.
22. Fisher and O'Gorman, *Women Combat Pilots*, DVD.
23. Pateman and Wyall, *We Were WASP*, VHS.
24. Gruhzit-Hoyt, *They Also Served*, 179.
25. WASP on the Web, "Arnold — Archive — Arnold letter," http://www.wingsacrossamerica.us (accessed October 17, 2009).
26. Laurel Ladevich, Mary McDonnell, and David G. McCullough, *American Experience: Fly Girls*, DVD (Boston: WGBH, 2006).
27. H.H. Arnold, *Global Mission* (New York: Arno, 1949), 358–359.
28. Fisher and O'Gorman, *Women Combat Pilots*, DVD.
29. Clark, *Dear Mother and Daddy*, 189.
30. Molly Merryman, *Clipped Wings: The Rise and Fall of the Women Airforce Service Pilots (WASPs) of World War II* (New York: New York University Press, 1998), 182.
31. Cochran, *Stars at Noon*, 135.
32. Ibid., 133–134.
33. Keil, *Those Wonderful Women*, 354.
34. Ibid., 355–357.
35. Ibid., 386.
36. Greene and Buss, *Flying High*, 37.
37. Strebe, *Flying for Her Country*, 67.
38. Clark, *Dear Mother and Daddy*, 187.
39. Ibid., 190.
40. Ibid., 191.
41. Churchill, *On Wings to War*, 161.
42. Scharr, *Sisters in the Sky*, vol. 2, p. 715.
43. Ann Baumgartner Carl, *A WASP Among Eagles: A Woman Military Test Pilot in World War II* (Washington, D.C.: Smithsonian Institution Press, 1999), 104.
44. Pennington, *Wings, Women and War*, 143.
45. Ibid., 10.
46. *WASPs and Witches*, DVD.
47. Pennington, *Wings, Women and War*, 1.
48. Anne Noggle, *A Dance with Death: Soviet Airwomen in World War II* (College Station: Texas A & M University Press, 1994), x.
49. Pennington, *Wings, Women and War*, 161.
50. Ibid., 143.
51. Noggle, *Dance with Death*, x.
52. Cheesman, *Brief Glory*, 6.
53. Curtis, *Forgotten Pilots*, 281.
54. du Cros, *ATA Girl*, 87.
55. Curtis, *Forgotten Pilots*, 283.
56. Whittell, *Spitfire Women*, 220.
57. Moggridge, *Woman Pilot*, 116.
58. Curtis, *Forgotten Pilots*, xi.
59. Ibid., 275.
60. Churchill, *On Wings to War*, 64.
61. National WASP World War II Museum, "Nancy Love Biography," http://waspmuseum.org/nancy-love-biography (accessed February 11, 2008).

62. Rich, *Jackie Cochran*, 141.
63. WASP on the Web, "NPR broadcast of *All Things Considered* on July 3, 2009," http://www.wingsacrossamerica.us (accessed October 17, 2009).
64. Henry Saikada, *Heroines of the Soviet Union 1941–45* (Oxford, UK: Osprey, 2003), 6–7.
65. Cottam, *Women in War*, xxiv.
66. Pennington, *Wings, Women and War*, 212–213.
67. Ibid., 85.
68. Cottam, *In the Sky*, 99.
69. Pennington, *Wings, Women and War*, 104.
70. Yegorova, *Red Sky*, 207.
71. Fahie, *Harvest of Memories*, 182–183.
72. Ibid., 161.
73. Cheesman, *Brief Glory*, 218.

Chapter 8

1. Verges, *On Silver Wings*, 247.
2. Carl, *WASP Among Eagles*, 111.
3. Cole, *Women Pilots*, 139–140.
4. Ringenberg, *Girls Can't Be Pilots*, 155–156.
5. Ibid., 295–296.
6. Strohfus, *Love at First Flight*, 63.
7. Cole, *Women Pilots*, 147.
8. Ibid., 145.
9. Clark, *Dear Mother and Daddy*, 196.
10. Cole, *Women Pilots*, 148.
11. Ibid., 149.
12. Fagan, *Zoot Suits and Parachutes*, 191.
13. Turner, *Out of the Blue*, 456.
14. Greene and Buss, *Flying High*, 181.
15. Ibid., 179.
16. Ibid., 186.
17. Gruhzit-Hoyt, *They Also Served*, 258.
18. Churchill, *On Wings to War*, 171.
19. Hawkins, *Hazel Jane Raines*, 219.
20. Cottam, *In the Sky*, 188.
21. Pennington, *Wings, Women and War*, 208.
22. Cottam, *In the Sky*, 83.
23. Cottam, *Women in War*, xv, 10.
24. Ibid., 74.
25. Strebe, *Flying for Her Country*, 80–81.
26. Moggridge, *Woman Pilot*, 140.
27. *WASPs and Witches*, DVD.
28. Walker, *Spreading My Wings*, 214.
29. Volkersz, *The Sky and I*, 198–199.
30. Ibid., 198.
31. Ibid., 196.
32. Ibid., 199.
33. Ibid., 199–200.
34. du Cros, *ATA Girl*, 93.
35. Ibid., 85.
36. Escott, *Women in Air Force Blue*, 290.
37. Render, *No Place for a Lady*, 141.
38. Strebe, *Flying for Her Country*, 76.
39. Ibid., xi.
40. Merryman, *Clipped Wings*, 144–149.
41. Clark, *Dear Mother and Daddy*, 193.
42. Vera S. Williams, *WASPs: Women Airforce Service Pilots of World War II* (Osceola, WI: Motorbooks International, 1994), 138.
43. Strebe, *Flying for Her Country*, 77.
44. Merryman, *Clipped Wings*, 154.
45. Granger, *On Final Approach*, iii.
46. Ibid., 56.
47. Strebe, *Flying for Her Country*, 77–78.
48. Scharr, *Sisters in the Sky*, vol. 1, p. 508.
49. Gruhzit-Hoyt, *They Also Served*, 179.
50. Strebe, *Flying for Her Country*, 78.
51. Ibid., 78–79
52. Ibid., 79.
53. Fisher and O'Gorman. *Women Combat Pilots*, DVD.
54. David Luff, *Amy Johnson, "Enigma in the Sky": An Official Biography* (Shrewsbury, UK: Airlife, 2002), 324.
55. Midge Gillies, *Amy Johnson: Queen of the Air* (London: Phoenix, 2004), 412.
56. Luff, *Amy Johnson*, 342.
57. du Cros, *ATA Girl*, 64–65.
58. *WASPs and Witches*, DVD.
59. Cottam, *Women in War*, 152.
60. Pennington, *Wings, Women and War*, 141.
61. Strebe, *Flying for Her Country*, 27.
62. Scharr, *Sisters in the Sky*, vol. 1, p. 380.
63. Simbeck, *Daughter of the Air*, 244.
64. Gott, *Hazel Ah Ying Lee*, 35.
65. Ibid., 44.
66. Ibid., 67–68.
67. Ibid., 66.
68. Bartels, *Sharpie*, 286.
69. Linda Grant De Pauw, *Battle Cries and Lullabies: Women in War from Prehistory to the Present* (Norman: University of Oklahoma Press, 1998), 262.
70. Walker, *Spreading My Wings*, 77.
71. Gruhzit-Hoyt, *They Also Served*, 188.
72. Babington Smith, *Amy Johnson*, 378.
73. "Americans in the British ATA," MS 3920.
74. Strebe, *Flying for Her Country*, 82.
75. Cottam, *Women in War*, 39.
76. Pennington, *Wings, Women and War*, 3; Cottam, *Women in War*, 3.
77. Keyssar and Pozner, *Remembering War*, 39.
78. Cottam, *In the Sky*, 139.
79. Fisher and O'Gorman, *Women Combat Pilots*, DVD.
80. Ibid.
81. Ibid.
82. Tiburzi Bonnie, *Takeoff!: The Story of America's First Woman Pilot for a Major Airline* (New York: Crown, 1984), 166.
83. Ibid., 256.

84. Magid, *Women of Courage*, VHS.
85. Cole, *Women Pilots*, 102–103.
86. Cottam, *In the Sky*, 213.
87. Pennington, *Wings, Women and War*, 169.
88. Churchill, *On Wings to War*, 161.
89. Walker, *Spreading My Wings*, 194.
90. du Cros, *ATA Girl*, 68.

Chapter 9

1. Rich, *Jackie Cochran*, 2.
2. Ibid., 111.
3. Ibid., 86.
4. Ibid., 182.
5. Ibid., 104.
6. Scharr, *Sisters in the Sky*, vol. 2, pp. 456–457.
7. Rich, *Jackie Cochran*, 140.
8. Ibid., 141.
9. Brinley, *Jackie Cochran*, 244–245.
10. Cochran, *Stars at Noon*, vi.
11. Gower, *Women with Wings*, 41.
12. Carl, *WASP Among Eagles* , 87.
13. Gower, *Women with Wings*, 215.
14. Fahie, *Harvest of Memories*, 75.
15. Ibid., 177–178.
16. Whittell, *Spitfire Women*, 93.
17. Fahie, *Harvest of Memories*, 187, 190, 198.
18. Sakaida, *Heroines of the Soviet Union*, 15.
19. Ibid.
20. Ibid., 17.
21. Noggle, *Dance with Death*, 28.
22. Cottam, *Women in Air War*, 10.
23. Cottam, *Women in War*, 27.
24. Noggle, *Dance with Death*, 153.
25. Cottam, *Women in Air War*, 12.
26. Cottam, *Women in War*, 28.
27. National WASP World War II Museum, "Nancy Love Biography," http://www.waspmuseum.org (accessed September 24, 2008).
28. Rickman, *Nancy Love*, 49.
29. Ibid., 57.
30. Ibid., 73.
31. Ibid., 237.
32. Ibid., 110.
33. Ibid., 115.
34. Ibid., 220.
35. Ibid., 228.
36. Ibid., 272.
37. Haynsworth and Toomey, *Amelia Earhart's Daughters*, 301.

Chapter 10

1. Curtis, *Lettice Curtis*, 22.
2. Ibid., 40.
3. Ibid., 60.
4. Ibid., 51.
5. Ibid., 52.
6. Ibid., 76.
7. Ibid., 183.
8. Ibid., 87.
9. Ibid., 101.
10. Ibid., 106.
11. Ibid., 118.
12. Ibid., 127.
13. Curtis, *Forgotten Pilots*, 19.
14. Curtis, *Lettice Curtis*, 218.
15. *MailOnline*, s.v. "Lettice Curtis," http://www.dailymail.co.uk/news/article-516816 (accessed January 21, 2010).
16. Knapp, *New Wings for Women*, 94.
17. Render, *No Place for a Lady*, 69.
18. Ibid., 68.
19. Ibid., 77.
20. Knapp, *New Wings for Women*, 95.
21. Render, *No Place for a Lady*, 90.
22. Ibid., 136–137.
23. Knapp, *New Wings for Women*, 98.
24. Render, *No Place for a Lady*, 136, 138.
25. Babington Smith, *Amy Johnson*, 132.
26. Ibid., 139–140.
27. Ibid., 166.
28. Ibid., 173.
29. Ibid., 185.
30. Gillies, *Amy Johnson*, 113.
31. Babington Smith, *Amy Johnson*, 292–293.
32. Gillies, *Amy Johnson*, 276.
33. Babington Smith, *Amy Johnson*, 140.
34. Ibid., 366.
35. Fahie, *Harvest of Memories*, 157.
36. Moggridge, *Woman Pilot*, 17–18.
37. Ibid., 36.
38. Ibid., 42.
39. Ibid., 63–66.
40. Ibid., 77.
41. Ibid., 123.
42. Ibid., 140.
43. Martlesham Heath Aviation Society, "An obituary — Jackie Moggridge (1922–2004)," http://www.mhas.org.uk/MHAS/runway22/200405.pdf (accessed January 21, 2010).
44. Cottam, *Women in War*, 3.
45. Knapp, *New Wings for Women*, 133.
46. Cottam, *Women in War*, 5.
47. Ibid.
48. Cottam, *Women in War*, 6.
49. Ibid., 7.
50. Ibid., 8.
51. Ibid., 9.
52. Pennington, *Wings, Women and War*, 109.
53. Ibid., 136.
54. Cottam, *In the Sky*, 225.
55. Strebe, *Flying for Her Country*, 28.
56. Saikada, *Heroines of the Soviet Union*, 14–15.
57. Ibid., 15.
58. Strebe, *Flying for Her Country*, 27.
59. Yegorova, *Red Sky*, xxvii.

60. Ibid., 15.
61. Ibid., 40.
62. Ibid., 56.
63. Ibid., xxvii.
64. Ibid., 71.
65. Ibid., 77.
66. Ibid., 107.
67. Saikada, *Heroines of the Soviet Union*, 19–20.
68. Yegorova, *Red Sky*, 156.
69. Ibid., 208.
70. Carl, *WASP Among Eagles*, 21.
71. Ibid., 24.
72. Ibid., 35.
73. Ibid., 52.
74. Ibid., 2, 96.
75. Turner, *Out of the Blue*, 137.
76. Churchill, *On Wings to War*, xi.
77. Ibid., 6.
78. Ibid., 8.
79. Ibid., 18, 23.
80. Ibid., 32.
81. Ibid., 46.
82. Fisher and O'Gorman, *Women Combat Pilots*, DVD.
83. Churchill, *On Wings to War*, 92.
84. "WAFS, James, T.," MSS 250, TWU.
85. Churchill, *On Wings to War*, 165.
86. Ibid., 131.
87. Ibid., xi.
88. Ibid., 151.
89. Ibid., 163.
90. Ibid., xi.
91. Ibid., 168.
92. Reese et al., *Brief Flight*, VHS.
93. Ibid.
94. Ibid.
95. Gott, *Hazel Ah Ying Lee*, 96.
96. Reese et al., *Brief Flight*, VHS.
97. Gott, *Hazel Ah Ying Lee*, 97.
98. Ibid., 1.
99. Reese et al., *Brief Flight*, VHS.
100. Gott, *Hazel Ah Ying Lee*, 86.
101. Ibid., 35.
102. Reese et al., *Brief Flight*, VHS.
103. Kerfoot, *Propeller Annie*, 1.
104. Ibid., 39.
105. Ibid., 43.
106. Ibid., 44.
107. Ibid., 57–59.
108. Ibid., 60.
109. Ibid., 67.
110. Ibid., 62.
111. Ibid., 64.
112. Ibid., 66.
113. Ibid., 83.
114. Ibid., 84.
115. Whittell, *Spitfire Women*, 207.
116. Kerfoot, *Propeller Annie*, 92.
117. Ibid., 94.
118. Turner, *Out of the Blue*, 134.
119. Kerfoot, *Propeller Annie*, 95.
120. Ibid., 71.
121. Ibid., 78–79.
122. Ibid., 97.
123. Churchill, *On Wings to War*, 126.
124. Kerfoot, *Propeller Annie*, 99.
125. Scharr, *Sisters in the Sky*, vol. 1, p. 36.
126. Ibid., 28.
127. Ibid., vol. 2, p. 182.
128. Ibid., vol. 1, p. 234.
129. Ibid., vol. 2, p. 719.
130. Ibid., 722.
131. Ibid., 740.
132. Ibid., 745.
133. Bartels, *Sharpie*, 40.
134. Ibid., 61.
135. Ibid., 75.
136. Ibid., 82.
137. Ibid., 88.
138. Ibid., 113.
139. Ibid., 166.
140. Ibid., 178.
141. Ibid., 184.
142. Ibid., 200.
143. Ibid., 218.
144. Ibid., 225.
145. Ibid., 240.
146. Ibid., 241.
147. Ibid., 247.
148. Ibid., 244.
149. Ibid., 254.
150. *WASPs and Witches*, DVD.
151. Judy Lomax, *Flying for the Fatherland: The Century's Greatest Pilot* (New York: Bantam, 1988), 29.
152. Hanna Reitsch, *Flying Is My Life*, trans. Lawrence Wilson (New York: Putnam's, 1954), 103.
153. Ann Welch, *Happy to Fly: An Autobiography* (London: John Murray, 198), 21.
154. Lomax, *Flying for the Fatherland*, 46.
155. "Reitsch, Hanna," three biofiles, TWU.
156. Lomax, *Flying for the Fatherland*, 57.
157. Ibid., 181.
158. Reitsch, *Flying Is My Life*, 151.
159. Lomax, *Flying for the Fatherland*, 84.
160. Reitsch, *Flying Is My Life*, 187.
161. Lomax, *Flying for the Fatherland*, 103.
162. Reitsch, *Flying Is My Life*, 211.
163. Ibid., 128.
164. Ibid., 137.
165. Ibid., 154.
166. Ibid., 202.
167. Ibid., 239.
168. Lomax, *Flying for the Fatherland*, 53.
169. Hoffmann, *Stauffenberg*, 94.
170. Ibid., 95.
171. Ibid., 280.

Bibliography

Books and Media

Adams, Nelson, Katherine King, and Marjorie Margolies. *Silver Wings and Santiago Blue*. VHS. Washington, D.C.: PBS Video, 1980.

Arnold, H.H. *Global Mission*. New York: Arno, 1949.

Babington Smith, Constance. *Amy Johnson*. London: Collins, 1967.

Bartels, Diane Ruth Armour. *Sharpie: The Life Story of Evelyn Sharp, Nebraska's Aviatrix*. Lincoln, NE: Dageforde, 1996.

Bird, Nancy. *My God! It's a Woman*. Sydney, Australia: Angus & Robertson, 1990.

Bragg, Janet Harmon. *Soaring Above Setbacks: The Autobiography of Janet Harmon Bragg, African American Aviator*. Washington, D.C.: Smithsonian Institution Press, 1996.

Brinley, Maryann Bucknam. *Jackie Cochran: An Autobiography*. New York: Bantam, 1987.

Brontman, Lazar, and L. Khvat. *The Heroic Flight of the "Rodina."* Moscow: Foreign Languages, 1938.

Carl, Ann Baumgartner. *A WASP Among Eagles: A Woman Military Test Pilot in World War II*. Washington, D.C.: Smithsonian Institution Press, 1999.

Cheesman, E.C. *Brief Glory: The Story of A.T.A.* Leicester, UK: Harborough, 1946.

Churchill, Jan. *On Wings to War: Teresa James, Aviator*. Manhattan, KS: Sunflower University Press, 1992.

Clark, Marie Mountain. *Dear Mother and Daddy: World War II Letters Home from a WASP, an Autobiography*. Livonia, MI: First Page, 2005.

Cochran, Jacqueline. *The Stars at Noon*. Boston: Little, Brown, 1954.

Cole, Jean Hascall. *Women Pilots of World War II*. Salt Lake City: University of Utah Press, 1992.

Conant, Jennet. *The Irregulars: Roald Dahl and the British Spy Ring in Wartime Washington*. New York: Simon & Schuster, 2008.

Cooper, Ann L. *How High She Flies: Dorothy Swain Lewis, WASP of WWII, Horsewoman, Artist, Teacher*. Arlington Heights, IL: Aviatrix, 1999.

Cottam, Kazimiera Janina. *In the Sky Above the Front: A Collection of Memoirs of Soviet Airwomen, Participants in the Great Patriotic War*. Manhattan, KS: MA/AH, 1984.

_____. *Women in Air War: The Eastern Front of World War II*. Nepean, Canada: New Military, 1997.

_____. *Women in War and Resistance: Selected Biographies of Soviet Women Soldiers*. Nepean, Canada: New Military, 1998.

Curtis, Lettice. *The Forgotten Pilots: A Story of the Air Transport Auxiliary 1939–45*. Henley-on-Thames, Oxfordshire, UK: G.T. Foulis, 1971.

_____. *Lettice Curtis: Her Autobiography.* Walton on Thames, Surrey, UK: Red Kite, 2004.

De Pauw, Linda Grant. *Battle Cries and Lullabies: Women in War from Prehistory to the Present.* Norman: University of Oklahoma Press, 1998.

du Cros, Rosemary. *ATA Girl: Memoirs of a Wartime Ferry Pilot.* London: Frederick Muller, 1983.

Escott, Beryl E. *The WAAF: A History of the Women's Auxiliary Air Force in the Second World War.* Buckinghamshire, UK: Shire, 2003.

_____. *Women in Air Force Blue: The Story of Women in the Royal Air Force from 1918 to the Present Day.* Northamptonshire, UK: Patrick Stephens, 1989.

Fagan, O. Vivian. *Zoot Suits and Parachutes: WASPs WWII.* Hawaii[?]: O.V. Fagan, 1999.

Fahie, Michael. *A Harvest of Memories: The Life of Pauline Gower, M.B.E.* Peterborough, UK: GMS, 1995.

Ferry, Doris. *Western Memories: History and Recollections of the Western Australian WAAAF 1941–1946.* Woodvale, Western Australia: Doris Ferry, 1996.

Fisher, Jack, and Robert O'Gorman. *Women Combat Pilots: The Right Stuff.* DVD. New York: A & E Television Networks, 2003.

Gillies, Midge. *Amy Johnson: Queen of the Air.* London: Phoenix, 2004.

Gott, Kay. *Hazel Ah Ying Lee: Women Airforce Service Pilot, World War II, a Portrait.* McKinleyville, CA: Kay Gott, 1996.

_____. *Women in Pursuit: Flying Fighters for the Air Transport Command Ferrying Division during World War II, a Collection & Recollection.* McKinleyville, CA: Kay Gott, 1993.

Gower, Pauline. *Women with Wings.* London: John Long, 1938.

Granger, Byrd Howell. *On Final Approach: The Women Airforce Service Pilots of W.W.II.* Scottsdale, AZ: Falconer, 1991.

Greene, Betty, and Dietrich Buss. *Flying High: The Amazing Story of Betty Greene and the Early Years of Mission Aviation Fellowship.* Camp Hill, PA: Christian, 2002.

Gruhzit-Hoyt, Olga. *They Also Served: American Women in World War II.* New York: Birch Lane, 1995.

Hardesty, Von, and Dominick Pisano. *Black Wings: The American Black in Aviation.* Washington, D.C.: National Air and Space Museum, 1983.

Hawkins, Regina Trice. *Hazel Jane Raines: Pioneer Lady of Flight.* Macon, GA: Mercer University Press, 1996.

Haynsworth, Leslie, and David Toomey. *Amelia Earhart's Daughters: The Wild and Glorious Story of American Women Aviators from World War II to the Dawn of the Space Age.* New York: William Morrow, 1998.

Hoffmann, Peter. *Stauffenberg: A Family History, 1905–1944.* Montreal & Kingston: McGill-Queen's University Press, 1995.

Keil, Sally Van Wagenen. *Those Wonderful Women in Their Flying Machines: The Unknown Heroines of World War II.* New York: Rawson, Wade, 1979.

Kerfoot, Glenn. *Propeller Annie: The Story of Helen Richey, the Real First Lady of the Airlines.* Lexington: Kentucky Aviation History Roundtable, 1988.

Keyssar, Helene, and Vladimir Pozner. *Remembering War: A U.S–Soviet Dialogue.* New York: Oxford University Press, 1990.

Knapp, Sally. *New Wings for Women.* New York: Thomas Y. Crowell, 1946.

Ladevich, Laurel, Mary McDonnell, and David G. McCullough. *American Experience, Fly Girls.* DVD. Boston: WGBH, 2006.

Langley, Wanda. *Flying Higher: The Women Airforce Service Pilots of World*

War II. North Haven, CT: Linnet, 2002.

Lomax, Judy. *Flying for the Fatherland: The Century's Greatest Pilot*. New York: Bantam, 1988.

Luff, David. *Amy Johnson, "Enigma in the Sky": An Official Biography*. Shrewsbury, UK: Airlife, 2002.

Magid, Ken. *Women of Courage: The Story of the Women Pilots of World War II*. VHS. Lakewood, CO: K.M. Productions, 1993.

Merryman, Molly. *Clipped Wings: The Rise and Fall of the Women Airforce Service Pilots (WASPs) of World War II*. New York: New York University Press, 1998.

Moggridge, Dolores Theresa. *Woman Pilot*. London: Michael Joseph, 1957.

Noggle, Anne. *A Dance with Death: Soviet Airwomen in World War II*. College Station: Texas A & M University Press, 1994.

Pateman, Yvonne, and Gene Wyall. *We Were WASP: Women Airforce Service Pilots, WWII*. VHS. Mission Viejo, CA: Video Movie Magic, 1990.

Pateman, Yvonne C. *Women Who Dared: American Female Test Pilots, Flight-test Engineers, and Astronauts, 1912–1996*. Norstahr, 1997.

Pennington, Reina. *Wings, Women and War: Soviet Airwomen in World War II Combat*. Lawrence: University Press of Kansas, 2001.

Pisano, Dominick A. *To Fill the Skies with Pilots: The Civilian Pilot Training Program, 1939–46*. Urbana and Chicago: University of Chicago Press, 1993.

Reese, Wayne, Alan H. Rosenburg, and Ming-Na. *A Brief Flight: Hazel Ying Lee and the Women Who Flew Pursuit*. VHS. Pacific Palisades, CA: LAWAS Productions, 2002.

Reitsch, Hanna. *Flying Is My Life*. Translated by Lawrence Wilson. New York: Putnam's, 1954.

Render, Shirley. *No Place for a Lady: The Story of Canadian Women Pilots, 1928–1992*. Winnipeg, Manitoba, Canada: Portage & Main, 1992.

Rich, Doris. *Jackie Cochran: Pilot in the Fastest Lane*. Gainesville: University Press of Florida, 2007.

Rickman, Sarah Byrn. *Nancy Love and the WASP Ferry Pilots of World War II*. Denton: University of North Texas, 2008.

_____. *The Originals: The Women's Auxiliary Ferrying Squadron of World War II*. Sarasota, FL: Disc-Us, 2001.

Ringenberg, Margaret J., and Jane L. Roth. *Girls Can't Be Pilots: An Aerobiography*. Fort Wayne, IN: Daedalus, 1998.

Sakaida, Henry. *Heroines of the Soviet Union, 1941–45*. Oxford, UK: Osprey, 2003.

Scharr, Adela Riek. *Sisters in the Sky*. Vol. 1, *The WAFS*. Vol. 2, *The WASP*. St. Louis, MO: Patrice, 1986.

Simbeck, Rob. *Daughter of the Air: The Brief Soaring Life of Cornelia Fort*. New York: Atlantic Monthly Press, 1999.

Stalin, Joseph. *The Great Patriotic War of the Soviet Union*. New York: International, 1945.

Stevenson, Clare, and Honor Darling. *The WAAAF Book*. Sidney, NSW, Australia: Hale & Iremonger, 1984.

Strebe, Amy Goodpaster. *Flying for Her Country: The American and Soviet Women Military Pilots of World War II*. Westport, CT: Praeger Security International, 2007.

Strohfus, Elizabeth, and Cheryl J. Young. *Love at First Flight: One Woman's Experience as a WASP in World War II ... and Fifty Years Later, She's Still Flying*. St. Cloud, MN: North Star, 1994.

Tanner, Doris Brinker. *Zoot-Suits and Parachutes and Wings of Silver Too!: The World War II Air Force Training of Women Pilots, 1942–1944*. Paducah, KY: Turner, 1996.

Thomson, Joyce A. *The WAAAF in Wartime Australia*. Carlton, Victoria, Australia: Melbourne University Press, 1991.

Tiburzi, Bonnie. *Takeoff!: The Story of America's First Woman Pilot for a Major Airline*. New York: Crown, 1984.

Timofeyeva-Yegorova, Anna. *Red Sky, Black Death: A Soviet Woman Pilot's Memoir of the Eastern Front*. Bloomington, IN: Slavica, 2009.

Turner, Betty Stagg. *Out of the Blue and into History*. Arlington Heights, IL: Aviatrix, 2001.

Verges, Marianne. *On Silver Wings: The Women Airforce Service Pilots of World War II, 1942–1944*. New York: Ballantine, 1991.

Volkersz, Veronica. *The Sky and I*. London: W.H. Allen, 1956.

Walker, Diana Barnato. *Spreading My Wings: One of Britain's Top Woman Pilots Tells Her Remarkable Story*. Somerset, UK: Patrick Stephens, 1994.

WASPs and Witches: The History of Women Pilots Fighting for the Right to Fight. DVD. Produced and directed by Jamie Doran. Atlantic Celtic Films, 2000.

Welch, Ann. *Happy to Fly: An Autobiography*. London: John Murray, 1983.

Whittell, Giles. *Spitfire Women of World War II*. London: HarperPress, 2007.

Williams, Vera S. *WASPs: Women Airforce Service Pilots of World War II*. Osceola, WI: Motorbooks International, 1994.

Journal Articles

Johnson, Ann R. "The WASP of World War II," *Aerospace Historian* 17, no. 2–3 (1970).

Milstead, Vi Warren. "Flying with the ATA: A Girl Air Transport Auxiliary Pilot Recalls Her Wartime Experiences." *CAHS Journal* (Fall 1999).

Manuscripts

"African American WASP Applicants." MSS 250, Blagg-Huey Library, Texas Woman's University, Denton.

Leveaux, Roberta Sandoz. "Americans in the British ATA." MSS 392c, Woman's Collection, Blagg-Huey Library, Texas Woman's University, Denton.

"Reitsch, Hanna." Three Bio-files, Woman's Collection, Blagg-Huey Library, Texas Woman's University, Denton.

"WAFS — James, T." MSS 250, Woman's Collection, Blagg-Huey Library, Texas Woman's University, Denton.

"WASP Biofiles, 44-W-9, Gee, Margaret (Maggie)." MSS 250, Woman's Collection, Blagg-Huey Library, Texas Woman's University, Denton.

Index

Numbers in ***bold italics*** indicate pages with photographs.

Agazarian, Monique 36
Air Transport Auxiliary (ATA) 3, 4, 8, 13, 21–23, 25–28, 33–38, 40, 41, 45, 48–55, 61, 63, 64, 67, 72, 77–81, 83–90, 92, 93, 96–99, 107, 111, 112, 114, 123, 129–131, 138, 139, 142, 147, 150, 156–160, 163, 164, 166, 180–182, 186
Air Transport Command 28–32, 35, 50, 64, 94, 116
Air Trips 12, *13*, 149, 150
Allen, Myrtle 40
Anderson, Charles Alfred 42
Arnold, Henry Harley "Hap" 19, 25, 26, 30–32, ***31***, 46, 65, 76, 81, 92, 94, 115–119, 124, 133, 147, 148, 155, 173
Arnold, W. Bruce 133
Aronova, Raisa Yermolayevna 138
Auriol, Jacqueline 130, 166, 192
Avenger (newspaper) 62
Avenger Field 3, 30 58, 59, 61, 62, 65, 78, 85, 96, 101, 178, 182
Aviation Group #122 8, 23, 45
awards 24, 111, 124, 125, 148, 150, 151, 154, 155, 162, 167, 169, 172, 176, 190, 192

Balfour, Harold 21
Barnett, Earsly Taylor 42
Batson (Crews), Nancy B. 47, ***75***, 187
Batten, Jean 38
Belik, Vera 108
Bendix Transcontinental Air Race 16, 146
Bespalova, Galina 109
Bird, Nancy 38, 76
Blinova, Klavdiya 106
Boylan, Margaret Kerr 116
Bragg, Janet Harmon Waterford 43
British Overseas Airways Corporation (BOAC) 22, 45, 54, 125, 150, 166
Brok, Galina 68
Brown, Willa 42
Budanova, Ekaterina "Katya" V. 169
Bureau of Air Commerce 17, 99, 181
Butler, Lois 36

Camp Davis 83, 90, 91, 102, ***103***, 173

Carl, Ann Baumgartner 78, 90, 91, 102, 121, 126, 172, 173, 182
Carpenter, Esther Nelson ***75***
Carter, Jimmy 134
Carter, Mildred Hemmons 43
Caterpillar Club 99
Central Airlines 179–181
Central Flying School 84
Channel Airways 166
Chapin, Emily 40
Chechneva, Marina Pavlovna 69
Chinese Air Force 176, 177
Cholmondeley, Victoria 39
Civil Aeronautics Administration 19, 26, 71, 181
Civil Air Guard 18, 21, 34, 72, 130, 137, 163
Civil Air Patrol 34, 42, 57, 172, 174
Civilian Pilot Training Program (CPTP) 19, 20, 32, 35, 89, 116, 172, 182, 186
Clark, Helen Mary ***75***, 174
Clark, Marie Mountain 46, 47, 59, 62, 70, 71, 75, 78, 91, 92, 99, 100, 117, 118, 121, 127, 132
Cochran, Jacqueline 3–5, 8, 16, 17, 25, ***27***, 29–32, 35, 40, 42, 43, 45–47, 50, 52, 54, 57–62, ***60***, 65, 70, 73–76, 78, 81, 83, 84, 90, 94, 97, 98, 102, ***103***, 114–120, 124, 128, 130, 132, 145–148, ***148***, 154, 157, 160, 166, 173, 176, 181, 182, 186, 192
Coffey School of Aeronautics 42
Cole, Jean Hascall 100, 101, 126, 141
Costello (WASP) Bill 116–119
Costello, John M. 116
Cousins, Rose Rolls 43
Crossley, Winnifred 23
Cunnison, Margaret 23
Curtis, Lettice 39, 50, 63, 66, 67, 87, 111, 112, 123, 156–159

Dahl, Roald 66
Daily Mail 88, 162, 163
Deaton, Leni Leoti "Dedie" Clark 52, 60–63
D'Erlanger, Gerard 22, 63, 72, 150
Dickerson, Patricia "Pat" A. 179
Disney, Walt 33, 66

Index

Dive (Day) Bomber Regiment 24, 25, 68, 106; *see also* 587th Aviation Regiment; 125th Guards Aviation Regiment
DuCros, Rosemary Rees 23, 33, 51, 66, 67, 84–86, 88, 112, 123, 131, 135, 142
Duffy, Dorothy 59
Duffy, Helen 59, 75
Duhalde, Margot 39, 54
Dulaney, Gini 128

Earhart, Amelia 15, 16, 23, 42, 43, 146, 160, 172, 180, 181
Edinger, Babette DeMoe 92
Eisenhower, Dwight 148
Elizabeth Bowes-Lyon, Queen consort 80
Engels 45, 55–57, 64, 68, 109, 168
Englund, Irene 134
Erickson, Elizabeth 101
Erickson (London), Barbara 81, 90, 124, 134, 138, 140, 186
Everard-Steenkamp, Rosamund — 39
Everleigh, Peggy 34

Fagan, Opal Vivian Hicks 34, 47, 59, 63, 71, 78, 101, 102, 128
Fairweather, Margaret "Margie" 18, 23, 158
Ferry Pilots Notes 87, 157
Fifinella 65, 66
Fifinella Gazette 62, 74
Fighter Regiment 24, 105; *see also* 586th Aviation Regiment
588th Aviation Regiment 24, 104, 105, 107–109, 112, 125, 138
587th Aviation Regiment 24, 25, 34, 68, 104, 106, 107, 109, 125, 152
586th Aviation Regiment 24, 50, 56, 68, 69, 72, 73, 104–106, 109, 112, 125, 129, 169
Fort, Cornelia 35, 70, 81, 136
46th Guards Aviation Regiment 19, 50, 56, 105, 112, 125, 138; *see also* 588th Aviation Regiment
46th Taman Guards Aviation Regiment 125; *see also* 588th Aviation Regiment; 46th Guards Aviation Regiment
Friedlander, Mona 23
Furey, Dorothy 28

Gee, Margaret "Maggie" 44
Gelman, Polina 141
George, Harold L. 29, 30, 35
George VI 80
Gething, Mardi 39
Gillies, Betty Huyler 75, 76, 81, 155
Golbinec, Leona 101
Goldwater, Barry 133
Gore, Margot Wyndham 23, 77, 87, 125
Gorman (Graba), Gretchen 136
Gott (Chaffey), Lois Kay 75, 89, 136, 179
Gower, Jennie 59
Gower (Fahie), Pauline 4, 5, 8, 11–14, *13*, 17, 21, 22, 39–41, 48, 50, 51, 63, 66, 72, 79, 80, 84, 115, 123, 125, 130, 145, 149–151, 157, 158, 161, 163
Granger, Byrd Howell 49, 74, 76, 92, 94, 100, 103, 116, 117, 133, 140
Greene, Elizabeth "Betty" 65, 90, 91, 102, 120, 128, 173
Gridnev, Aleksandr 73, 106
Grizodubova, Valentina S. 23, 34, 41, 69, 124, 129, 151, 166–168
Gwinn aircar 17, 153

Hamble (ferry pool) 23, 66, 77, 123
Harris, Arthur 26
Harrison, Helen 19, 39, 52, 63, 85, 92, 131, 159, 160
Hartson, Mary 112
Haslemere (ship) 164
Hatfield (ferry pool) 22, 23, 33, 39, 54, 77, 79, 80, 157, 164, 165
Hawson, Mary 101
Haydu, Bernice "Bee" Falk 118, 133
Henderson, Nellie 141
Hinckley, Robert H. 19, 20
Hitler, Adolf 5, 8, 18, 39, 188–192
Hobby, Oveta Culp 45, 76, 116
Hughes, Joan 23
Hultgreen, Kara 139, 140
Humphreys, Jack 161, 162

James, Teresa D. 57, 58, 64, *75*, 92, 96, 121, 129, 141, 174–176, *175*
Jason *162*
Johnson, Amy 21, 38, 72, 80, 134, 135, 138, 157, 160–164, *163*
Johnson, Sadie Lee 43
Johnson, Sylvia Dalmes 179

Kazarinova, Militsiia A. 152
Kazarinova, Tamara Aleksandrovna 73, 105, 106
Keil, Mary Ellen 101
Khomiakova, Valeriia 73
King George VI 80
Komsomol 18, 24, 36, 51, 169
Krivonogova, Sasha 34, 129

Ladies Courageous (film) 175, 186
Large, Gloria 38
Law, Ruth 11
Lee, Hazel "Ah Ying" 44, 89, 136, 176–179, *178*
Lend-Lease 25, 28, 29, 85, 89, 94, 109, 177
Leska, Anna 39, 54
LeValley, Gertrude Meserve (Tubbs) *75*
Leveaux, Roberta Sandoz 40, 55, 81, 86, 87, 138
Lewis, Dorothy "Dot" M. Swain 58, 59, 91
Littlewood, Margaret 38
Litvyak, Lidya "Lilya (Lily)" 64, 69, 73, 135, 168, 169
Lindberg, Charles 189

Lomanova, Galina Tenuyeva 152
Look (magazine) 175
Love, Nancy Harkness 5, 8, 16, 17, 28, 29, 32, 35, 47, 48, 50, 52, 57, 58, 64, 73, 74, 76, 78, 81, *82*, 97, 114, 116, 119, 124, 145, 147, 153–155, *154*, 174, 181, 183, 184, 186
Love, Robert 28, 124, 153
Lyle, Nancy 7

Macon Telegraph 40
Madison, Isabel 136
Makarova, Tat'yana 19, 108
Markov, Valentin 106, 107
Marsalis, Frances Harrell 179
Martin, Peggy 112
McElroy, Lenore 81
McGilvery, Helen *75*
McIntyre, Dorothy Lane 43
Miles, Maxine "Blossom" 21
Miller (Watson), Florene 47, 71, 79, 90, 100, 127
Milstead, Violet Warren 38, 55, 63, 86, 87, 131
Moggridge, Dolores Theresa "Jackie" Sorour 39, 40, 51, 54, 67, 123, 130, 165, 166
Mollison, James 162
Monserud, Nels 70, 71
Mosdale (ship) 40

National Air Marking Program 17, 99, 181
National League of Pilots 18
New York Times 29, 52, 146, 173
Nicholas, Betty Pettitt 129
Nicholson, Mary 40, 112
Night Bomber Regiment 24, 57, 68, 107; *see also* 588th Aviation Regiment; 46th Guards Aviation Regiment; 46th Taman Guards Aviation Regiment
night witches 4, 39, 56, 107, 127
Ninety-Nines 42, 128, 147, 185
Nkrumah, Kwame 192
Norbeck, Jeanne L. 112
Nyman, Marylene "Geri" Lamphere 138

Obama, Barack 124
O'Connor, Ann Cawley 102
Odlum, Floyd B. 16, 146, 147
Officer Candidate School (OCS) 59, 117, 118, 147
Oldenberg, Margaret 46
Olds, Robert 29, 153
Omlie, Phoebe 58, 59, 83
125th Guards Aviation Regiment 50, 104, 110, 112, 125, 129; *see also* 587th Aviation Regiment
Orr, Marion 38, 99
Osadze, Irina 129
Osipenko, Polina 24, 124, 151, 167, 168
Osoaviakhim 34, 129, 166, 170

Parker, Mary 182

Pateman, Yvonne C. "Pat" 90, 112, 182
Patterson, Gabrielle 23
Pearl Harbor (attack) 33, 35, 44
Pearson, Drew 116, 147
phoney war 50, 164
Pilsudska, Jadwiga 39
Pittman, Bessie *see* Cochran, Jacqueline
Prokhorova, Evgeniia "Zhenia" 104, 105
Protasova, Svetlana 130

Queen consort 80

race 41–44, 94
Raines, Hazel J. 39, 40, 78, 84, 86, 90, 129
Ramspeck, Robert 116
Rangitata (ship) 39
Raskova, Marina 4, 8, 14, 15, *15*, 17, 19, 23, 24, 34, 36, 41, 45, 50, 51, 55–57, 64, 69, 72, 73, 77, 105–107, 109, 115, 124, 125, 141, 145, 151–153, 167, 168, 171
Ratcliffe (ferry pool) 157
Rawlinson, Mabel 83, 102
Reitsch, Hanna 4, 5, 7, 18, 156, 187–192, *191*
Rhonie (Brooks), Aline "Pat" 174, 183
Richey, Helen 28, 49, 90, 92, 98, 140, 179–182, *180*
Ringenberg, Margaret Ray 70, 78, 112, 121, 126, 127
Robinson, Marie Michell 112, 113
Rodina 4, 14–16, 23, 41, 115, 124, 129, 167
Roosevelt, Franklin D. 19, 26
Roosevelt, Eleanor 20, 25, 28, 42, 80, 181
Roulstone (Reeves), Frances R. 37, 96, 101
Royal Air Force Volunteer Reserve 130, 166
Royal Canadian Air Force 3, 37, 159
Rozanova, Larisa Litvinova 108
Rudneva, Zhenya 139
Russell (Burnett), Elspeth 38, 131

Schaefer, Helen M. 136
Scharr, Adela Riek 43, 47, 71, 76, 90, 91, 94, 95, 117, 121, 133, 136, 147, 182–186, *183*
Schiller, Melitta, Countess von Stauffenberg 5, 7, 192, 193
Scott, Betty Mae 112
Scott, Dorothy 47
Seafarer 162
Sharp, Evelyn 35, 46, 47, 76, 137, 185–187, *187*
Sheehy, Ethel 43, 173
Shelmerdine, Francis 22
Shiel, Maureen E. 125
Skoblikova, Antonina 129
Slack, Autumn Geneva *178*
Slinn, Dorothy Fulton *75*
Solomatina, Zinaida Fedorovna 129
Spicer, Dorothy 12, 14, 21, 149, 150
Stag Lane 12, 21, 79, 161
Stalin, Joseph 5, 18, 23, 24, 49, 51, 69, 73, 105, 110, 111, 122, 152, 167
statistics 38, 49, 50, 59, 105, 107, 111–113, 120, 126

Stegeman, Marion 182
Stinson, Katherine 11
Strodl (Dowling), Vera 96, 131, 132
Strohfus, Elizabeth 34, 46, 47, 61, 71, 72, 90, 117, 127
Strother, Dora Dougherty 133
Studilina, Nadya 108
Sullivan, Madeline 127

Tanner, Doris Brinker 62
Thompkins, Gertrude "Tommy" 113
Tibbets, Paul W. 94
Tiburzi, Bonnie 140
Tomassy, Fran Snyder 179
Treaty of Versailles 18, 189
Truman, Harry S. 42
Tunner, William H. 45, 89, 94
Turabelidze, Galina 69
Tuskegee Airmen 42

Udet, Ernst 25, 189
uniforms *15*, 53, 59, 60, 63–66, 74, *75*, *103*, 141, *148*, *154*, *175*, 177, *178*, 182, *191*
Urban, Robert R. 65
Urban's turbans 65

Volkersz, Veronica 33, 48, 54, 67, 87, 98, 130, 131
Von Greim, Ritter 190, 191
von Stauffenberg, Alexander 5
von Stauffenberg, Claus 5

Walker, Diana Barnato 36, 54, 77, 84, 86, 88, 97, 137
Warren (Doerr), Virginia Lee 38
Welch, Ann 189
Westover, Oscar 25
Whirly Girls 189
Whitchurch (ferry pool) 67
White Waltham (ferry pool) 28, 40, 54, 77, 80, 123, 157, 158
Wilberforce, Marion 23
Wojtulanis, Stefania "Barbara" 39, 54

Women Airforce Service Pilots (WASP) 2, 3, 5, 8, 20, 31, *31*, 32, 35, 37, 38, 42–44, 46, 47–50, 52, 58–62, *60*, 65, 66, 70, 71, 74–76, 78, 89–92, 94, 97, 98, 101–103, *103*, 107, 109, 112, 114–121, 124, 126, 127, 132–137, 139, 142, 147, *148*, 154, 155, 173, 176–187, *178*, *180*, *183*
Women's Air Training Corps (WATC) 38
Women's Army Auxiliary Corps (WAAC) 35, 42, 45
Women's Army Corps (WAC) 43, 46, 70, 76, 115, 116
Women's Auxiliary Air Force (WAAF) 33, 34, 67, 165
Women's Auxiliary Australian Air Force (WAAAF) 3, 38, 39
Women's Auxiliary Ferrying Squadron (WAFS) 3, 5, 8, 20, 28, 29, 30, 31, 32, 35, 37, 45, 47–50, 57, 58, 64, 73–76, *75*, 78, 81, *82*, 90, 97, 98, 112, 116, 120, 124, 126, 136, 138, 142, *154*, *175*, 176, *183*, 184, 186
Women's Flying Training Detachment (WFTD) 8, 30–32, 37, 43, 52, 57, 58, 62, 65, 73, 74, 78, 98, 142
Women's Royal Air Force (WRAF) 131
Wood, Ann 145, 157
Wood, Betty Taylor 102
Work, Robert 191
Works Progress Administration 28
Wright, Orville 173

Yakovleva, Olga 56, 57
Yamshchikova, Olga Nikolayevna 129
Yeager, Charles "Chuck" 148
Yegorova, Anna Timofeyeva 5, 110, 111, 125, 169–172
Yeremenko, Sasha 129

Zapol'nova, Yevgeniya 104
Zelenko, Yekaterina 111
Zhigulenko, Andreyevna 57, 68, 107, 108, 112, 139
Zillner, Lorraine 101

www.ingramcontent.com/pod-product-compliance
Ingram Content Group UK Ltd.
Pitfield, Milton Keynes, MK11 3LW, UK
UKHW041959140426
5217IPUK00015B/874